Spotlights on Contemporary Family Life

FAMILYPLATFORM
Families in Europe Volume 2

Edited by Linden Farrer & William Lay

This publication was produced by FAMILYPLATFORM, 2011.

ISBN 978-1-4475-1660-6.

FAMILYPLATFORM (SSH-2009-3.2.2 Social platform on research for families and family policies) is funded by the EU's 7th Framework Programme (€1,400,000) for 18 months (October 2009-March 2011).

The consortium consists of the following 12 organisations:

1. Technical University Dortmund (Co-ordinators)
2. State Institute for Family Research, University of Bamberg
3. Family Research Centre, University of Jyväskylä
4. Austrian Institute for Family Studies, University of Vienna
5. Demographic Research Institute, Budapest
6. Institute of Social Sciences, University of Lisbon
7. Department of Sociology and Social research, University of Milan-Bicocca
8. Institute of International and Social Studies, Tallinn University
9. London School of Economics
10. Confederation of Family Organisations in the European Union (COFACE), Brussels
11. Forum Delle Associazioni Familiari, Italy
12. MMMEurope (Mouvement Mondial des Mères-Europe), Brussels

Contact *info@familyplatform.eu* or visit *http://www.familyplatform.eu* for more information.

Thanks to ILGA-Europe for use of photo on the front cover.

Typesetting and front and back cover graphic design by Lila Hunnisett (*http://lilahunnisett.com/*).

This document is produced and distributed under a *Creative Commons Attribution-NonCommercial-NoDerivs 3.0 Unported* licence. This allows copying, distribution and transmission, with the condition that it is properly attributed, used for non-commercial purposes, and that no derivative works are created. The full legal code is available at: *http://creativecommons.org/licenses/by-nc-nd/3.0/legalcode.*

EUROPEAN COMMISSION
European Research Area

SEVENTH FRAMEWORK PROGRAMME

Funded under Socio-economic Sciences & Humanities

Unless otherwise stated, the views expressed in this publication do not necessarily reflect the views of the European Commission.

Contents

Preface by Linden Farrer and William Lay 5

Chapter 1: Structures & Forms 9

    Editorial
    Epp Reiska 9
1.1 Towards a Definition of the Family?
    Leeni Hansson 11
1.2 Major Trends in Family Behaviour in European Countries
    Leeni Hansson 21
1.3 Changes in Finnish Families:
    Towards Full-Time Motherhood and a New Familialism?
    Marjo Kuronen, Teppo Kröger and Kimmo Jokinen 33
1.4 Short Account of Changes in the Family in Italy
    Carmen Leccardi and Miriam Perego 36
1.5 Trends in the German Family Model: Pluralisation of Living
    Arrangements, and Decrease in the Middle-Class Nuclear Family
    Ursula Adam, Loreen Beier, Dirk Hofaecker, Elisa Marchese, Marina Rupp 39
1.6 Families in Hungary
    Zsuzsa Blaskó 43

Chapter 2: Solidarities in Contemporary Families 45

    Editorial
    Carmen Leccardi and Miriam Perego 45
2.1 How Social Change is Transforming Relations Between
    the Generations
    Interview with Claudine Attias-Donfut 48
2.2 Family Solidarity and the New Forms of Social Uncertainty
    Interview with Carla Facchini and Marita Rampazi 59
2.3 Ambivalences, Conflicts and Solidarities Within the Family Today
    Interview with Ariela Lowenstein 74
2.4 Intergenerational Solidarity and EU Citizens' Opinions:
    Some Indications for Policy Making
    Francesco Belletti 84
2.5 Intergenerational Solidarity:
    Rebuilding the Texture of Cities
    Lorenza Rebuzzini 99
2.6 Solidarity in Large European Families
    Raul Sanchez 110

Chapter 3: Demographic Change and the Family in Europe 113

Editorial
Veronika Herche 113

3.1 **Demographic Changes and Challenges in Europe**
Interview with Paul Demeny 116

3.2 **Are Babies Making a Comeback?**
Interview with Professor Herwig Birg 125

3.3 **Family Changes in the New EU Member States**
Zsolt Spéder 133

3.4 **Childbearing in a Gender-Equal Society**
Interview with Livia Sz. Oláh 150

3.5 **"To be or not to be and how to fill empty cradles? That is the question"**
Zsuzsanna Kormosné-Debreceni 161

Chapter 4: Volunteering in Families 167

Editorial
Anne-Claire de Liedekerke, Joan Stevens and Julie de Bergeyck 167

4.1 **Volunteering in the European Union: An Overview of National Differences in the EU Member States**
Birgit Sittermann 169

4.2 **Volunteers in the EU Spotlight: The European Year of Volunteering 2011**
John MacDonald and Sara Lesina 176

4.3 **Demographic Change and Its Impact on Voluntary Work**
Christiane Dienel 185

4.4 **Family and Education Towards Voluntary Action**
Francesco Belletti and Lorenza Rebuzzini 191

4.5 **Towards the European Policy Agenda on Volunteering: Taking Into Account the Needs of Families**
Edited by Gabriella Civico 196

4.6 **Volunteering and Service in the United States**
Barb Quaintance 205

Contributors 211

Preface

Linden Farrer and William Lay
The Confederation of Family Organisations in the European Union

Spotlights on Contemporary Family Life combines the four volumes of the *FAMILYPLATFORM Online Journal*. Covering issues of cross-cutting importance to families, it is one of the outcomes of FAMILYPLATFORM, a 'social platform' working together to chart and review the major trends of family research in the European Union, critically review existing research, attempt to foresee future challenges facing families, and to bring all of this work together in the form of a research agenda on families for the European Union. The diagram below sketches out the steps taken towards realisation of the research agenda[1].

Diagram: Outline of key steps to the European Research Agenda

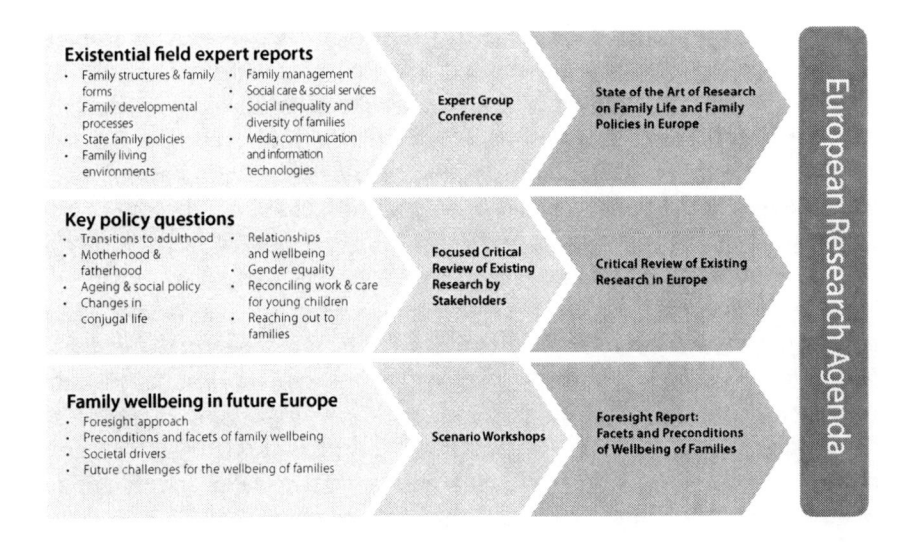

Existential field expert reports
- Family structures & family forms
- Family developmental processes
- State family policies
- Family living environments
- Family management
- Social care & social services
- Social inequality and diversity of families
- Media, communication and information technologies

→ Expert Group Conference → State of the Art of Research on Family Life and Family Policies in Europe

Key policy questions
- Transitions to adulthood
- Motherhood & fatherhood
- Ageing & social policy
- Changes in conjugal life
- Relationships and wellbeing
- Gender equality
- Reconciling work & care for young children
- Reaching out to families

→ Focused Critical Review of Existing Research by Stakeholders → Critical Review of Existing Research in Europe

Family wellbeing in future Europe
- Foresight approach
- Preconditions and facets of family wellbeing
- Societal drivers
- Future challenges for the wellbeing of families

→ Scenario Workshops → Foresight Report: Facets and Preconditions of Wellbeing of Families

→ European Research Agenda

Source: *FAMILYPLATFORM (2011)*.

Although the main outcome of the project is the European Research Agenda, none of the steps leading up to this - the summary of the major trends of families in the "Existential Fields" of family life, the critical review of existing research, and the 'future of families' exercise - would have been possible without the active participation of representatives of policy, scientific and social organisations from across

---

[1] Final reports of all of the main steps are available from the FAMILYPLATFORM website (*http://www.familyplatform.eu*). All of these reports will be available in *Family Wellbeing: Challenges for research and policy* (Uhlendorff, Rupp & Euteneuer, 2011).

Europe. As representatives of family organisations, the Confederation of Family Organisations in the European Union (COFACE) - alongside Mouvement Mondial des Mères - Europe (MMMEurope) and Forum delle Associazioni Familiari - helped galvanise the support of different stakeholders, and ensure that the grassroots concerns of families were properly represented in FAMILYPLATFORM. The idea was to enable the input of stakeholders at every stage, and this included the three Info Days (organised in Budapest, Milan and Brussels), the two conferences taking place in Lisbon and Brussels, and online on the interactive website.

In addition to raising the profile of FAMILYPLATFORM and encouraging participation of stakeholders, one of the main responsibilities was publication of the *FAMILYPLATFORM Online Journal*. Four volumes were to be produced, but neither the subject matter nor the format had been defined. Early on, roles were made clear: COFACE would manage editing and dissemination of the volumes, and the editors of each of the four volumes would be responsible for agreeing the subject matter and soliciting contributions. This arrangement was first tested by Leeni Hansson and Epp Reiska (University of Tallinn), and apart from a few changes that gave a greater voice to policy makers and civil society, the model continued without change for the next three volumes.

When it came to deciding on content, one thing was very clear early on: certain subjects are of cross-cutting relevance to families in Europe, and connect the different fields of family life (the so-called "Existential Fields") and the different policy questions discussed by participants during the course of the project. For this reason, the Online Journal was an opportunity to discuss important subjects that might otherwise be overlooked.

With this in mind, *Chapter 1: Structures and Forms (previously Volume 1)* presents an overview of issues relating to the structure and the form of families in Europe today. From an academic perspective, the issues are quite clear: the characteristics of different forms, the numbers and proportions of people living in different family forms, the causes and consequences of the emergence of new family models. For policy makers and family associations the issues are more complicated. Many family organisations do not represent all family forms, and the debate can get quite heated on what constitutes a family and what effect different family forms have on the wellbeing of members of families, particularly children. It is for this reason that FAMILYPLATFORM agreed early on not to define the family, and concentrate instead on different aspects of family life. This seemed the best way of avoiding intractable debates and excluding voices from publicly funded research. Needless to say, *Chapter 1* presents a thorough overview of key trends across Europe, and illustrates some of the differences found between different parts of Europe: a reminder that no 'one model fits all' family policy is currently applicable or relevant to all of the countries of the European Union.

Chapter 2: Solidarities in Contemporary Families saw Carmen Leccardi and Miriam Perego (University of Milan-Bicocca) take up the editorial baton, this time to describe the shifting solidarities found within European families. Too often the issue of an 'ageing society' is simply reduced to the problem of over-burdening social care systems – but longevity also represents opportunities for new kinds of solidarities inside families and family networks, and new relations between family members. As such, the subject covered in *Chapter 2* cross-cuts with an issue high on the European policy agenda: intergenerational solidarity. It is therefore no surprise that 2012 is currently designated the European Year of Active Ageing, though it is hoped that the title will soon include the words "intergenerational solidarity" as well.

Next in line was Veronika Herche at the Demographic Research Institute, who edited *Chapter 3: Demographic Change and the Family in Europe*. With articles giving voice to stakeholders with quite different opinions on the issues at stake and the opportunities and challenge they raise, this chapter opens up a lot of areas for further discussion – areas that are high on the policy agenda, this time being a priority of the (current) Hungarian Presidency of the Council of the European Union.

The final chapter, *Chapter 4: Volunteering and the Family*, was edited by colleagues at MMMEurope. It would have been an oddity if this issue had strayed from issues of importance to European policy, and this issue was no exception. Inspired by the MMMEurope survey of mothers, and by family associations who argue that families play a big role in volunteering, this chapter gives an overview of what's known - and what isn't - about volunteering by families. Coinciding with the European Year of Volunteering 2011, it is a timely look at the efforts that families put into volunteering across Europe and the important benefits that Europe gains from all of this voluntary effort.

Although the four volumes were originally only meant to be published electronically, they were deemed worthy of being printed and becoming a more permanent feature in the literature on European families. In addition to the many reports of FAMILYPLATFORM, the different brochures and booklets, and of course the European Research Agenda which is in many ways a distillation of all of the different messages heard whilst working on this project, we hope that *Spotlights on Contemporary Family Life* raises awareness of important issues, and helps pave the way for family policies that work towards improving the wellbeing of families in Europe, both today and tomorrow.

Chapter 1: Structures & Forms

Editorial

Epp Reiska
The Institute for International and Social Studies of Tallinn University, Estonia

This is the first chapter of *Spolights on Contemporary Family Life*. The overall objective of the FAMILYPLATFORM project is to elaborate a research agenda that addresses fundamental research issues and key policy questions for future research and family policies in Europe. The aim of the project is to improve the well-being of families by understanding future challenges families in Europe will face. FAMILYPLATFORM is not a research project: it addresses research issues and policy questions by reviewing existing research, identifying significant trends and differences between countries, and by exposing research gaps and problems relating to methodology.

This chapter, being the first, focuses on broad - but nevertheless - important issues relating to research on families. The first paper addresses the difficulties of trying to define the family. In everyday life "family" is a commonly used word and everyone seems to understand what the word means. However, the second half of the twentieth century has witnessed major changes in family formation and family behaviour that have resulted in a diversification of family forms. Because of this, it has become more and more difficult to use a general and universally acceptable definition of the family. Concepts and definitions of the family have changed over time, and are used differently in policy formation and in academic literature. The article by Leeni Hansson Towards a Definition of the Family traces changes in the definition of the family both in academic literature as well as in policy formation.

The other purpose of this first chapter of the journal is to give an overview of the major trends in family behaviour in different European countries. In the article Major Trends in Family Behaviour in European Countries, Leeni Hansson addresses issues such as the child leaving the parental home, the choice between marriage and cohabitation, having children, family break-up and "living apart together". Based on those developments three visions for the future of families are highlighted.

Leeni's article is followed by short insights on family issues in four European countries that represent different paths and outcomes in the development of the family – Finland, Germany, Italy and Hungary. The article by Marjo Kuronen, Teppo Kröger and Kimmo Jokinen from Finland centres mainly on the drawbacks of the allegedly woman-friendly welfare state, where extensive public day-care is provided, offering possibilities to combine family life and paid work. Ursula Adam, Loreen Beier, Dirk Hofaecker, Elisa Marchese and Marina Rupp describe the incidence of the so-called "middle-class nuclear family", and the developments leading to this concept in Germany. Carmen Leccardi and Miriam Perego from Italy write about changes in the prevalence of marriage and within the couple itself towards a more equal relationship of women and men. Last but not least, developments in the Hungarian family are described by Zsuzsa Blaskó with reference to other European countries.

We hope that the contributions presented in the first chapter of our journal bring to your attention the complexity of issues related to families, and provide you with a better understanding of the similarities and differences in the developments in the institution of the family in different parts of Europe, which is ultimately the aim of our project.

1.1 Towards a Definition of the Family?

Leeni Hansson, Institute of International and Social Studies

The family is one of the basic social institutions or, as Goode (1964: 4) put it "the only social institution which is formally developed in all societies". In western culture, the family was traditionally defined in terms of a married couple with children, who shared a common home and divided family-related tasks and responsibilities along gender lines (Strong/DeVault, 1993). Despite major changes in many societies that have had a significant impact on family formation and family behaviour, the institution of the family retains its social importance. However, due to these changes in patterns of family formation and the diversification of family forms, it has become increasingly difficult to find a general and universally accepted definition of the family.

In everyday life "family" is a commonly used word, and everyone seems to understand what the word means. However, John Peters (1999: 55) has stated that "the term 'family' is one of the most misused concepts in the English language" because the word reflects a wide variety of social relationships. Furthermore, concepts and definitions of the family have changed over time.

The goal of this paper is to trace how the definition of the family has changed during the course of the second half of the twentieth century, both in academic literature and in terms of policy formation.

Structure and functions of the family

According to Merriam-Webster's Collegiate Dictionary (1993), the word "family" derives from the Latin word "familia" that originally meant household, and included the householder and his family members as well as kin and servants. In contemporary usage, the word "family" is most often used to refer to at least two different types of relationships: (1) to related people who live together in the same household - most often a husband, his wife and their child(ren), or (2) to a larger circle of persons – a network of relatives and kin not necessarily confined to one household.

Sociologists focus on two key issues when dealing with the family – the structure of the family, and its functions. The structure refers to the composition of family, i.e. to family members, their positions in the family, such as mother, father, son, daughter, grandfather, grandmother, and to family organisation, i.e. the set of rules that govern patterns of interaction within the family. We can differentiate two main structural types of families: nuclear families and extended families. A nuclear family is composed of two generations, parents and their offspring, while different extended family types are composed of at least three generations, for example parents, their children, grandparents, etc.

The functions of the family refer to the common and essential responsibilities that families fulfil both for society as well as for individual family members. For example, a family universally provides food and shelter, nurture and intimacy, and strategies for managing conflicts for its members. One of the most important functions of the family is the socialisation of children. This process includes raising and educating children, and familiarising them with the traditions and value systems of the community and culture they belong to.

Each society assigns specific roles to the family members. For example, up until the middle of the twentieth century, western countries were characterised by stereotypical attitudes towards family roles: it was assumed that the father was the main provider and instrumental task leader, and the mother was the main carer and homemaker (Scanzoni, 2001). Even today we can find unwritten social norms regarding the 'proper' roles of family members.

Definitions of the family used by social scientists

The definition of the family in western countries relied on three cornerstones - marriage, sex, and childbearing - up till the 1950s (Allan/Hawker/Crow, 2001). Marriage between a man and a woman was considered the foundation of the family and the only acceptable way of forming a new family. The ideal family type was a nuclear family headed by a man who was permanently married to his wife with the couple living with their common children. Another essential element of the family was a common dwelling place.

The best known definition of the family used in the mid-twentieth century was presented by American anthropologist George Peter Murdock in his study on social structure (Murdock, 1949). Murdock studied a sample of 250 different societies. Based on the results of his study, Murdock came to the conclusion that some forms of family existed in every society, and there was a common pattern that made it possible to formulate a definition of the family. According to Murdock's definition the family was as follows:

> "a social group characterized by common residence, economic cooperation, and reproduction. It includes adults of both sexes, at least two of whom maintain a socially approved sexual relationship, and one or more children, own or adopted, of the sexually cohabiting adults"
> (Murdock, 1949: 1-2).

Thus, the definition of the family formulated by Murdock was heteronormative, and characteristic first of all of the marriage-based nuclear family. According to Murdock's definition, a single mother with her children or a cohabiting couple, whose sexual relationship was not socially approved in the late 1940s, did not fit into the concept of the family. Murdock also identified what he believed to be the four main functions of the family: sexual, economic, reproductive and educational. These functions were supposed to reproduce the values and norms of the culture and community, and pass these values on to the next generation.

Talcott Parsons (1955) viewed the family as a sub-system within society, with its nature determined by its functions. Similar to Murdock, Parsons defined the family as a unit consisting of a married couple who co-operate in rearing children and who share a common place of residence. Parsons identified two essential functions of the family – socialisation of children and 'stabilisation of adult personalities'. Parsons also differentiated between gendered parental roles – the father's instrumental role on one side, and the mother's expressive role in the family on the other. Although there are debates as to how prevalent this family type actually was, it was given the label of 'traditional' family (Popenoe, 1987) and later the 'standard' family (Scanzoni, 2001).

In the 1960s, Europe experienced far-reaching demographic changes, mostly interpreted as responses to major economic developments, changes in the labour market and increasing female employment, and changes in attitudes towards gender roles. Across Europe, fertility and marriage rates declined and divorce, remarriage and unmarried cohabitation became accepted elements of family life. However, the definitions of the family used in the first half of the 1960s remained 'traditional'. For example, similar to Parsons, Coser (1964, cited in Peters, 1999) presents a definition of the traditional family as:

> "A group manifesting the following organizational attributes: It finds its origin in marriage; it consists of husband, wife, and children born in their wedlock, though other relatives may find their place close to this nuclear group, and the group is united by moral, legal, economic, religious and social rights and obligations (including sexual rights and prohibitions as well as such socially patterned feelings as attraction, piety, and awe"
> (Coser, 1964: xvi cited in Peters, 1999: 56).

Besides changes in demographic behaviour, the 1960s were also characterised by increasing tolerance and permissiveness in family life (Jallinoja, 1994). The cornerstones of the family of the 1950s - marriage and sex - were replaced by permissive attitudes towards interpersonal relationships and sex. As a result, marriage was no longer considered a lifelong commitment, and the 1960s were characterised by a considerable increase in rates of divorce and remarriage. Increased permissiveness brought about alternative lifestyle choices like non-marital cohabitation and extra-marital births. These radical changes in family behaviour made it increasingly difficult to define the family using the old touchstones that had formed the foundation of definitions in the 1950s. The model of the traditional family was no longer the only socially acceptable model because of the existence of a great variety of patterns 'hidden' behind this one concept (Liljeström, 2002).

In the 1960s and 1970s, the share of couples characterised by alternative lifestyle choices, i.e. non-married cohabiting couples with or without children, cohabiting homosexual couples, group marriages, communes, etc., increased in many western countries. However, even in the 1970s and 1980s, many family researchers still preferred to use traditional definitions of the family. For example in 1987 Eleanor Macklin still used the traditional definition of the family, similar to those used in the 1950s:

> *"a legal, lifelong, sexually exclusive marriage between one man and one woman, with children, where the male is primary provider and ultimate authority"*
> *(Macklin, 1987: 317).*

Soviet family sociology used a definition of the "proper" family, which was based on legal marriage, till the late 1980s; for example, the Dictionary of Sociology defined the family as:

> *"a social institution characterized by whole complex of social norms, sanctions, and modes of behaviour which regulate the relationship between the spouses and among parents and children' and 'a small group, founded on marriage or blood relationships, members of which are united by a common household"*
> *(Kratkii slovar…, 1988: 301 in Narusk, 1992).*

Thus, according to the definitions used by social scientists in the late 1980s, it was still assumed that the family should be based on heterosexual marriage, children, and a common place of residence. A divorced single mother with her children, cohabiting couples with or without children, married couples living in separate households because of geographically separate jobs, and homosexual couples did not qualify as families.

In the 1980s, alternative living arrangements increased in popularity in the majority of European countries. However, these new family forms and living arrangements were still perceived as non-standard or deviant compared to the traditional normative family, or not even families at all by policy makers (Scanzoni, 2001; Jallinoja, 1994).

However, while there were researchers who stuck to the definition of the normative family, there were also others who began to show a growing interest in the plurality of family forms and alternative lifestyle choices. General acceptance of diversity of family forms made it obvious that the definition of the family would expand. In the following statement, Rapoport and his colleagues characterised the changes that had taken place in the family:

> *"families of today are in a transition from coping in a society in which there was a single overriding norm of what family life should be like to a society in which a plurality of norms are recognized as legitimate and, indeed, desirable"*
> *(Rapoport/Fogarty/Rapoport, 1982).*

American sociologist Judith Stacey (1996) also described the changes that had taken place in the family as drifting away from the single dominant family model towards an increasing variety of family relationships. Stacey was among the first to state that gay and lesbian families - extremely diverse themselves - had a great role in developing the postmodern family. According to Stacey, it was high time to start speaking of the western family as the "postmodern family". Contrasted to the "traditional family", the postmodern family was characterised as ambivalent in terms of its gender and kinship arrangements.

By the 1980s, sociologists began to speak of a loosening of the strong marital bond between husband and wife, and of its diminishing significance as the primary family bond. Instead of the marital bond, parent-child relationships became the backbone of the family. Accordingly, based on the new family ideology, James White defined the family as:

> *"an intergenerational social group organized and governed by social norms regarding descent and affinity, reproduction of the young"*
> *(White, 1991: 37).*

If the common home of a married couple was one of the essential elements of definitions of the family in the 1950s, by the 1980s and 1990s it was not excluded even if family relations were dispersed over several households, and even over several cities or countries.

In European countries, the demographic changes that had had significant effects on the family took place at different times and at different rates (Hantrais/Letablier, 1996; Gauthier, 1996; Hantrais, 2005): there were countries where marriage was no longer a necessary precondition for setting up a family, and there were countries where divorce was not even acceptable in the 1990s. Whether cohabitation was considered a pro*per set*ting in which couples could have children or not also varied across countries. In Nordic countries, for example, cohabitation acquired an equal footing with marriage (Liljeström, 2002). Accordingly, it is not a surprise that the definitions of the family provided by Nordic sociologists give a considerably broader picture of the family than traditional definitions, as demonstrated by Norwegian family sociologist Arnlaug Leira:

> *"Usually, the term 'family' refers to at least two persons, either two adults who share bed and table, as a Norwegian expression goes, or one or more adults who take parental responsibility for one or more children. It may also refer to one or more adult child(ren) who share(s) a household with his/her parent(s)"*
> (Leira, 1996).

Thus, according to Leira's definition, different family forms can include males and females regardless of whether they are married or have children, and they can include women or men cohabiting with a partner of the same sex regardless of whether they have children or not, and even include people who live with siblings or non-related others.

In the 1990s, there was increasing recognition that the definition of the family needed to be more inclusive and less restrictive. Theresa Rothausen defined the family as:

> *"a group of people who are interested in one another due to dependence, obligations or duty, love, caring, or cooperation"*
> (Rothausen, 1999: 819).

Thus, in Rothausen's definition, the concept of the family is expanded to include individuals who are not necessarily tied by marriage and reproduction, but also those who are tied by love and caring functions.

In 1990, Jaber Gubrium and James Holstein published a book titled *What is Family?* that was deeply critical of traditional family sociology. The main point of the book was to reject the concept of "the family" and to begin to develop a concept of the process associated with 'being a family'. Gubrium and Holstein's ideas were later supported by several family sociologists. Diana Gittins (1993) suggested that due to major changes in family structure, it would be more appropriate to use the term "families" than "the family". Even today there are different opinions as to what the family is or should be.

The majority of definitions coming out of the late 1990s had one thing in common – concepts of the family were no longer strictly related to marriage. Indeed, Elisabeth Beck-Gernsheim (2002) pointed out that few institutions changed in the last decades of the twentieth century more than the family.

Definitions of the family in official use

The way the family is defined by the state determines what rights and responsibilities are recognised and expected by legal and other social institutions (Skolnick, 1981). European countries recognise the family as an important social institution and the majority of countries have adopted different policy measures designed to protect and support the family. However, whether these measures are targeted at one particular type of family or at all families depends on how the family is defined by the policy makers. For example, in countries with strong Roman Catholic beliefs and where marriage is still considered a defining feature of the family, some of the policies could easily be targeted at the 'standard' family only (Hantrais/Letablier, 1996).

Besides family policies and different social policies, official definitions of the family are needed to conduct population censuses. Over the course of the last few decades, definitions of the family used in censuses have also changed. For example, the definition of the family proposed by the United Nations in 1974 for use in population and housing censuses was based on marriage as the main defining criterion for the family (Hantrais, 2005). In UN recommendations for population and housing censuses published twenty years later, reference to marriage as a defining characteristic of a family was replaced by reference to 'cohabiting partners' (UNECE, 1998: 191).

Whilst acknowledging that significant changes have taken place in the institution of the family, it is easier for policy makers to target measures at the traditional marriage-based family. In only a few countries are non-marital cohabiting couples legally registered and considered to be families. Accordingly, Moen and Schorr (1987) proposed that rather than using a universal definition of the family, it would be more appropriate to define the family according to the particular issues involved. They suggested that when dealing with issues of child support the use of a definition including households with children is most appropriate, and when dealing with property settlements of cohabiting adults the use of a definition that includes intimate primary relationships would be more useful. Thus, in designing new policy measures, policy makers have to adjust to a diversity of family forms.

In 2002, the European Commission Directorate-General for Employment and Social Affairs published a research report from 15 EU member states, plus Iceland, Liechtenstein and Norway (European Commission, 2002), which revealed two categories of countries. First, there were countries which made a clear distinction between the concept of 'family' and 'household'. In these countries a sociological and legal value was attributed to the term 'family', and an economic value to the term 'household'. The group of countries with a clear distinction between the two concepts consisted of southern European countries plus the United Kingdom. The other group of countries either made no distinction between the concepts of 'family' and 'household', or used only the term 'household'. However, definition of the family - where it was used - was not universally the same. In some countries like Italy it was linked with marriage and in others it was not. Finally, there was the extreme case of Norway, where even an unmarried person living alone was considered to be a family (European Commission, 2002). The study revealed that in answer to the question on the role of the term 'family' in family benefits there were differences between countries that had used a precise definition of the term and those that had not. For example Denmark, Germany, Ireland, Iceland, Liechtenstein, Norway and the United Kingdom frequently used the term 'child' instead of 'family' when defining family benefits (European Commission, 2002).

In a majority of European countries, new forms of family (cohabiting couples, re-married couples, single-parent families, step-families, etc.) are entitled to benefits on an equal basis with marriage-based families. In this respect homosexual couples might be an exception. By the time the above mentioned report was published (2002), homosexual couples were legally recognised in France, the Netherlands, Norway, Finland and Sweden *(ibid.)*. Today, debates on the issue of homosexual families are in progress in several countries.

Conclusion

Arlene Skolnick has stressed that defining the family is not just an academic exercise, because the way it is defined determines which kind of families are considered normal and which are considered deviant (Skolnick, 1981). Although families continue to play a very important role in society and the nuclear family is still a normative ideal of family in many European countries, during the last few decades family structures have become highly diverse. New definitions of the family reveal that it is practically impossible to define the family using the touchstones of the family that characterised definitions used in the 1950s or 1960s.

When carrying out comparative studies on changes in family behaviour, and in the course of comparing family policies across governments, it is important to bear in mind that the definitions used may well not be the same. At the same time, these different definitions not only highlight different realities but also act as a window on the different values held by people across Europe towards the family.

References

- Allan, G., Hawker, S. & Crow, G. (2001). *Family diversity and change in Britain and Western Europe.* Journal of Family Issues 22.7: 819-837.
- Beck-Gernsheim, E. (2002). *Reinventing the Family: in search of new lifestyles.* Cambridge: Polity Press.
- European Commission (2002). *Family Benefits and Family Policies in Europe.* Directorate-General for Employment and Social Affairs, Unit E.2.
- Gauthier, A. H. (1996). *The State and the Family; A Comparative Analysis of Family Policies in Industrialized Countries.* Oxford: Clarendon Press.
- Gittins, D. (1993). *The Family in Question: Changing Households and Familiar Ideologies,* 2nd ed. Basingstoke: Palgrave Macmillan.
- Goode, W. J. (1964). *The Family.* Englewood Cliffs, NJ: Prentice Hall.
- Gubrium, J. F. & Holstein, J. A. (1990). *What is Family?* Mountain View, CA: Mayfield Publications.
- Hantrais, L. (2005). *Living as a family in Europe.* Paper presented at European Population Conference 'Demographic Challenge for Social Cohesion', 7-8 April 2005, Strasbourg.
- Hantrais, L. & Letablier, M.T. (1996). *Families and Family Policies in Europe.* London: Longman.
- Jallinoja, R. (1994). *Alternative family patterns; their lot in family sociology and in the life-worlds of ordinary people.* Innovation, 7.1: 15-27.
- *Kratkii slovar po sotsiologii [Short dictionary of sociology]* (1988). Moscow: Politizdat.
- Leira, A. (1996). *Parents, Children and the State: Family Obligations in Norway, Oslo.* Institute for Social Research, Report No. 96: 23.
- Liljeström, R. (2002). *The strongest bond on trial.* In Liljeström, R. & Özdalga, E. (eds.) *Autonomy and Dependence in the Family.* Istanbul: Swedish Research Institute.
- Macklin, E. (1987). *Non-traditional family forms.* In Sussman, M. & Steimetz, S. (eds.) *Handbook of Marriage and the Family.* New York: Plenum Press.
- *Merriam-Webster's Collegiate Dictionary* (10th edition) (1993). Springfield, MA: Merriam-Webster, Inc.
- Moen, P. & Schorr, A. L. (1987). *Families and Social Policy.* In Sussman, M. B. & Steinmetz, S. K. (eds.) *Handbook of Marriage and the Family.* New York: Plenum Press.
- Murdock, G. P. (1949). *Social Structure.* New York: The MacMillan Company.

- Narusk, A. (1992). *Parenthood, partnership, and family in Estonia.* In Björnberg, U. (ed.) *European Parents of the 1990s.* New Brunswick: Transaction Publishers.
- Parsons, T. (1955). *The American family: Its relations to personality and to the social structure,* In Parsons, T. & Bales, R. F. (eds.) *Family Socialization and Interaction Process.* Glencoe, IL: Free Press.
- Peters, J. F. (1999). *Redefining Western families.* In Settles, B.H., Steimetz, S.K., Peterson, G.W., Sussman, M.B. (eds.) *Concepts and Definitions of Family for the 21st Century.* Oxford: Haworth Press.
- Popenoe, D. (1987). *Beyond the Nuclear Family: A statistical Portrait of the Changing Family in Sweden.* Journal of Marriage and the Family 49.1: 173-183.
- Rapoport, R. N., Fogarty, M. P. & Rapoport, R. (1982). *Families in Britain.* London: Routledge & Kegan.
- Rothausen, T. (1999). *'Family' in organizational research: a review and comparison of definitions and measures.* Journal of Organizational Behavior 20: 817-830.
- Scanzoni, J. (2001). *From the Normal Family to Alternate Families to the Quest for Diversity With Interdependence.* Journal of Family Issues 22.6: 688-710.
- Skolnick, A. (1981). *The family and its discontents.* Society 18: 42-47.
- Stacey, J. (1996). *In the Name of the Family: Rethinking Family Values in the Postmodern Age.* New York: Basic Books.
- Strong, B. & DeVault, C. (1993). *Essentials of the Marriage and Family Experience.* St. Paul, MN: West Publishing Company.
- UNECE (1998). *Recommendations for the 2000 Censuses of Population and Housing in the ECE Region.* No 49. New York: United Nations.
- White, J. M. (1991). *Dynamics of Family Development: A Theoretical Perspective.* New York: The Guilford Press.

1.2 Major Trends in Family Behaviour in European Countries

Leeni Hansson
Institute of International and Social Studies, Tallinn University

In the majority of western countries, the second half of the twentieth century was characterised by significant changes in the family as an institution and an increasing plurality of family forms. The new form of family formation alongside families based on registered marriage - unregistered cohabitation - became an alternative to marriage-based families. Besides cohabitation, the share of other novel family structures – single-parent families, reconstituted families, living apart together (LAT), living together apart (LTA), etc., has also increased. Family sociologists attribute changes in the institution of family to a process of individualisation and de-institutionalisation of the family (Beck-Gernsheim, 2002) that is reflected in increasing variability of individual choices concerning the timing of family formation and the form of the union itself. The aim of this article is to provide an overview of the major changes that have occurred in the family in different EU countries.

Changes in family formation

Leaving parental home

Although there are differences in cultural attitudes and family policies, there is a trend for young people across Europe to remain in their parental home for longer than young people did some decades ago (Cherlin *et al.*, 1997; Corijn/Klijzing, 2001). According to a Eurobarometer survey carried out in 2005 (*Mobility in...*, 2006), young Europeans leave their parental home at the average age of 21, though this differs considerably across European countries. According to this survey, young people leave the parental home at a younger age in Nordic and Baltic countries (Denmark, Finland, Sweden, Estonia, Latvia, Lithuania), than in southern Europe *(ibid.)*. On average in 2008, half of the female population had left parental home by the age of 23, and half of male population by the age of 26 (*The Social Situation...*, 2010). It is characteristic of all the EU countries that women leave the parental home at a younger age than men. The general trend of remaining in the parental home for longer can be explained first of all by the increasing number of years spent in education and by postponement of marriage. Geographical differences in turn are mainly explained by the differences in opportunities to participate in the labour and housing markets.

Marriage-based families

Marriage was considered the primary foundation for family formation and a socially normative precondition for having children in western countries until the 1960s. In the second half of the twentieth century however, the situation changed, and since the 1970s the crude marriage rate, i.e. number of marriages per 1,000 population in the year, has decreased significantly in the majority of European countries (Kalmijn, 2007). According to Eurostat population data (*Europe in...*, 2010; *The Social Situation...*, 2010), in 1960 the crude rate of marriage was considerably lower in the 'old' EU countries (7.9) than in the 'new' member states (such as Latvia 11.0, Romania 10.7, Lithuania 10.1, and Estonia 10.0). The marriage rate was below the EU average in Nordic countries, but lowest in Ireland – 5.5 marriages per 1,000 of the population. According to population statistics *(ibid.)*, during the period between 1960–2007 the differences in marriage rates levelled, and in the countries where marriage rates had been high in the 1960s the decline was steepest. The steepest decline in marriage rates took place in the Baltic countries, which in the 1960s were characterised by relatively high marriage rates. For example, compared to Sweden where the crude marriage rate was already low in the 1960s, the decrease was from 6.7 in 1960 to 5.2 in 2007, whereas in Estonia the decrease was from 10.0 in 1960 to 5.2 in 2007, and in Latvia from 11.0 in 1960 to 6.8 respectively (*The Social Situation...*, 2010).

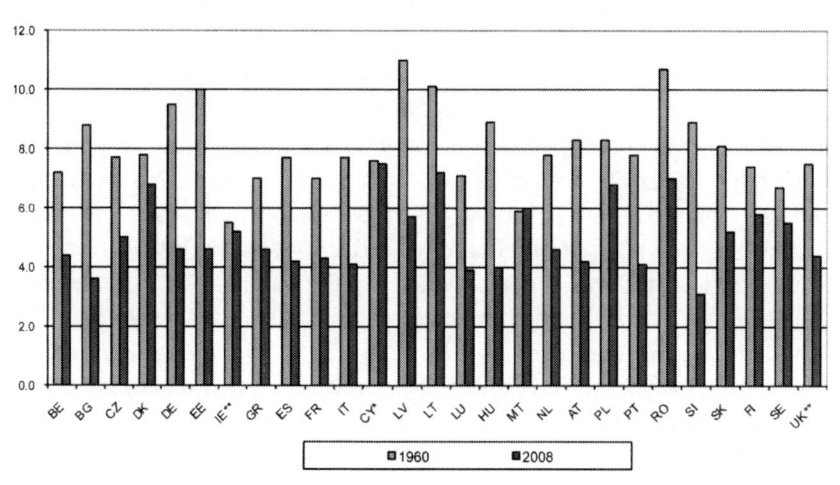

Figure 1. Marriages per 1,000 persons in 1960 and 2008 (EU27)

* Cyprus – data of 1970, data of 1960 not comparable.
Source: *Europe in Figures – Eurostat yearbook (2009)*.

One explanatory factor for the declining marriage rate is that family formation has increasingly become connected to alternative living arrangements, i.e. an increase in the share of cohabiting unions as an alternative to a marriage-based family (Kalmijn, 2007). The other explanatory factor is postponement of marriage. It is a general trend in the majority of European countries that the mean age of women at first marriage is increasing. In Northern Europe it has increased by almost four years during the past few decades. While in the EU15 the average age of women at marriage is 27.5 years, in northern countries it is close to 30 years. In Sweden a woman's average age at marriage increased from 24 years in 1960 to 30.2 years in 2000, in Denmark it increased from 22.8 to 29.5 years, and in Finland it increased from 23.8 to 28 years (*Population Statistics*, 2006).

It was characteristic of the former socialist countries that women got married on average two years earlier than women in western European countries. Although in both groups of countries the mean age of women at first marriage increased, by the turn of the century the differences between the EU15 and the 12 new member states remained about the same. Among new member states, the mean age of women at first marriage is highest in Slovenia (26.7), Malta (26.7) and Cyprus (26.4) – but they are all lagging behind the EU15 average. The only "old" EU country where the average age of women at first marriage is similar to the average of the new member states is Portugal (25.7); among the countries of EU15, it is the lowest age of a woman at first marriage.

Cohabiting unions

The crude rate of marriage is a statistical measure that takes into account only officially registered marriages and disregards other forms of partnership. Accordingly, in many countries, there are no relevant statistics concerning cohabitation available. Based on data contained in different surveys (e.g. *European Social Survey, Eurobarometer Survey, Fertility and Family Surveys*, etc.,) we can conclude that since the 1970s the popularity of living together without getting married has increased in the majority of European countries but the rapidity of the increase in cohabitation differs according to country (Brown & Booth, 1996). On the basis of these differences, several typologies of cohabitation have been constructed (Kasearu, 2007).

First, on the basis of the proportion of cohabiting unions, Kasearu *(ibid.)* divides European countries into three broad groups. In the first group are countries with a high proportion of cohabitating unions. The share of cohabitating unions among 26-35 year-old men and women is highest in Sweden, where 43% of people in that particular age group are cohabiting. Finland, Denmark, Norway and France are also characterised by relatively high levels of cohabiting unions, with around one in three individuals aged 26-35 in a cohabiting union. In the second group we find the United Kingdom, Belgium, Luxembourg and Estonia. In these countries, cohabitations constitute one in four unions; Netherlands, Germany, Austria, and Slovenia have cohabitation level of about 20%. The third group is characterised by the lowest levels of cohabitation, and these are seen in Greece, Portugal, Poland, Czech Republic, Slovakia, Hungary, Spain, Italy and Ireland, where the share of people in cohabiting unions is under 10%. The two EU countries that were not included in Kasearu's analysis - Bulgaria and Romania - appear to belong to the group of countries with low levels of cohabitation (see Hoem *et al.*, 2007). Therefore, it can be concluded that in many European countries marriage is losing its former popularity and cohabitation is increasing.

However, knowing the proportion of cohabiting couples does not provide us with a thorough overview of union-formation patterns and their causes in European countries. For example, in Portugal and Greece, the cohabitation level is low because the marriage rate is high: around two-thirds of women in their late twenties are already married, a fact that considerably lowers the proportion of cohabiting couples. The situation is different in Spain and Italy where women in their twenties are neither cohabiting nor married. This suggests that in some cases we cannot describe cohabitation as a substitute for marriage; however, in some other countries this explanation can be statistically proven (Kiernan, 2002).

In most countries, the share of cohabitation is age-dependent. This means that cohabitation is most popular among people in their twenties, and that the proportion of cohabiting unions declines with age (Kiernan, 2002). Cohabitation may have different meanings to different age groups. While for young people, living together without marriage may constitute a test period before marriage rather than an alternative to marriage, in older age groups cohabitation is often justified by external factors (people awaiting a divorce, adult children disapproving of the remarriage of their parents, reasons related to property, etc.).

Kiernan *(ibid.)* provides her own typology, which is constructed on the basis of the social acceptance of cohabitation. Acceptance is determined by childbearing outside marriage on the one hand, and through increases in the number of cohabiting couples on the other. A number of stages can be identified and are drawn from the experience of the Swedish population because Sweden has a long-standing tradition of non-marital cohabitation as a family form.

According to Kiernan, in the first stage cohabitation is chosen by only a small subgroup of a population as an alternative lifestyle choice; marriage is still overwhelmingly considered the foundation of the family. In the second stage, cohabitation becomes popular as a test period before marriage, though it is usually followed by marriage; this means that cohabitation is a short time period during which the quality of the relationship and the partner's personality is tested. Cohabiting partners usually have no children during this test period. In the third stage, cohabitation is a socially acceptable alternative to marriage and children are not rare in these relationships. Finally, cohabitation becomes an alternative form of family formation, cohabitation and marriage become equal, and the only thing that makes a difference is the existence or non-existence of the marriage certificate. According to Kiernan, Nordic countries have reached the fourth stage, and recent trends in eastern European countries suggest that some of the new member states, for example Estonia, are also making a transition to the fourth stage.

Families with children

The rapid decline in rates of birth caused concern throughout Europe in the early 1990s. Generally the countries with the most marked decrease in the birth rate were those where they used to be high, such as southern European and Baltic countries. Although the rate of decline was influenced to a certain extent by differences in population age composition, the real rate of decline was clear (*Recent Demographic…*, 2002).

According to Eurostat data (*The Social Situation…*, 2010), in the EU27 in 2007 the total fertility rate, i.e. the average number of children born to a woman, was 1.55, with the highest being found in Ireland (2.01). Rates above 1.8 children per woman were recorded in France, Sweden, Denmark and United Kingdom. Although in these countries the total fertility rate was considerably higher than the average in EU countries, it was still below replacement level (2.1). Countries with a critical level of births were Slovakia, where the total fertility rate was as low as 1.25, but also Romania (1.3), Poland (1.31), Portugal (1.33) and Italy (1.35). However, compared to the significant fall in fertility rates that took place in many countries in the first half of the 1990s, we can today speak of a certain recovery.

Several population surveys have revealed an interesting fact: in most European countries the actual number of children in the family is below the desired number of children (Cliquet/Avramov, 1998). Socio-economic factors (education, work, income), relational factors (age at start of childbearing, marital status), and biological-reproductive factors have been seen to influence the discrepancy between actual and intended number of children *(ibid.)*.

A recent, statistically important development in some countries with very low fertility levels is a substantial increase in the number of childless couples. According to the OECD family database (OECD Family Database,

2008, SF 7) in 2007 in the age group 30-34, the share of women without children was the highest in Luxembourg (42.9%) and lowest in Slovakia (9.2%). Part of the reason for the differences in the rates of childlessness is postponement of having children – characteristic of many EU countries. In 1960 in EU15 countries the mean age of a woman giving birth to her first child was 25.2. By 2007, the mean age was highest in Ireland (31.1) and lowest in Bulgaria (26.7) and Romania (27.9) (*The Social Situation...*, 2010). This trend of postponement of birth of the first child and a longer period of childlessness leaves young people more time for individual choices between the family and career, and can be seen as a sign of the de-institutionalisation of the family.

An increase in extra-marital births is one of the most characteristic changes in Europe in the decades analysed. In the EU15, the share of extra-marital births has increased from 5% in 1960 to 36% in 2007 (*Population Statistics*, 2008). The rate of extra-marital births has increased approximately evenly both in the 'old' and 'new' EU countries. However, there is a wide gap between the rates of extra-marital births in the different geographic regions, and this gap has widened over the decades. In northern countries the increase in the share of extra-marital births has increased considerably (in Sweden from 11% in 1960 to 55% in 2007; in Denmark from 8% to 46%, and in Finland from 4% to 41% respectively) though this increase has been much less marked in southern Europe, e.g. Greece (4.9), Italy and Spain, all three remaining below 10% in 2000.

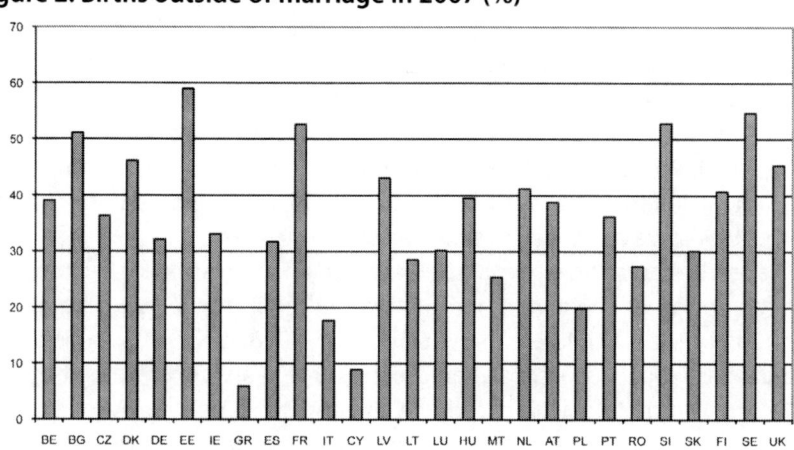

Figure 2. Births outside of marriage in 2007 (%)

* Ireland and Spain, data of 2006.
Source: *1960 - Eurostat: Population Statistics (2004:119).*
Source: *2008 - Eurostat: Europe in Figures - Eurostat Yearbook (2010:183).*

The 'new' EU countries reveal similar geographical patterns of extra-marital births: the rates are highest in Estonia (59%) and Latvia (43%) and lowest in Cyprus (*The Social Situation*..., 2010). It is worth mentioning that in Iceland, a non-EU country, 64% of all births were extra-marital in 2007 *(ibid.)*; the proportion of extra-marital births is also high in France (51%). The majority of extra-marital births are accounted to cohabiting couples, i.e. most children born outside marriage have a cohabiting mother and father (Lanciery, 2008).

Family break-up

The crude divorce rate[1] has increased over the past decades in most EU countries, except for Estonia and Latvia, where divorce rates were already high in the 1960s (*The Social Situation*..., 2010). In the 1960s, divorce rates were generally significantly higher in socialist block countries than in the western countries, but divorce rates increased faster in the 'old' EU countries, and differences between the country groups had levelled by 2000. According to Eurostat data (*Europe in Figures*, 2009), in 2007 among EU15 countries the crude divorce rate was high in Finland (2.5), Denmark (2.6), Spain (2.8) and Belgium (2.9). Among the new member states Lithuania (3.4) and Latvia (3.3) were characterised by the highest divorce rates followed by the Czech Republic (3.0) and Estonia (2.8). In contrast, Poland (1.7) and Slovenia (1.3) had the lowest number of divorces per 1,000 inhabitants among the new member states. There is no data for Malta as divorce is not legal there. Among the EU15, Greece (1.2) and Italy (0.8) had the lowest divorce rates in 2007 *(ibid.)*.

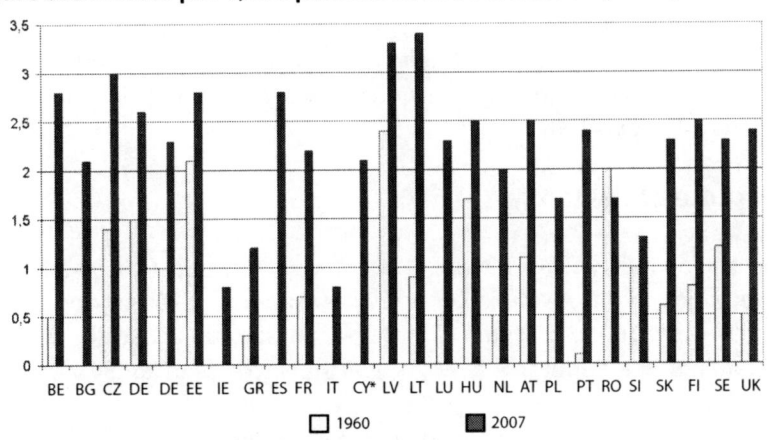

Figure 3. Divorces per 1,000 persons in 1960 and 2007 (EU27)

*France, data of 2006; no data for Malta as divorce is not legal.
Source: *1960 - Eurostat: Population Statistics (2004 :126)*.
Source: *2007 - Eurostat: Europe in Figures - Eurostat Yearbook (2010:184)*.

[1] Calculation based on the number of divorces in a given year per 1,000 inhabitants.

Because the dissolution of cohabiting relationships is not recorded, and non-marital cohabitation is increasing in the majority of European countries, it is very difficult to determine the real level of family break-ups.

Post-divorce families

In the 1960s, divorces were not widespread in western European countries, but the proportion of divorcees who re-married was considerably higher than today. While in the 1960s, 60-70% of divorcees (in Nordic countries 55%) re-married, by the end of the twentieth century this share had fallen to about 20%. Thus, today many divorced persons prefer not to remarry, and new permanent relationships are established as cohabiting unions instead.

Marital instability and an increase in separation and divorce have resulted in an increasing proportion of single parent families (Pryor/Rodgers, 2001). According to the OECD (*OECD Family database 2008*, SF 2), in 2007 the share of children aged 0-14 living in single-parent families was highest in United Kingdom (24%), followed by Estonia (18%) and Latvia (15%). The share of single-parent families was lowest in Romania (3%), Greece (4%), Italy and Malta (both 5%).

Living Apart Together (LAT)

The phenomenon of 'living apart together' (LAT) - where partners share common living arrangement for some periods of time, but also have separate residences (Trost, 1998) - is a relatively new living arrangement. LAT relationships are more common in Finland and Germany (for more detail, see *OECD Family database 2008*, SF 9). However, categorisation of the people involved as either single or partnered is problematic, because the situation can vary across countries and individual cases, and accordingly needs further study (Speder, 2007).

The future of the family

The increasing variety of family forms in recent years is the basis for several hypotheses regarding the future of the family (Cliquet/Avramov, 1998). In general, there are three main visions of the future of the family: (1) disappearance of the family; (2) restoration of the traditional family; (3) further increase in family variation.

The first scenario – disappearance of the family

Recent trends in various demographic indicators of family life, such as the decreasing popularity of marriage, instability of the family, decreasing fertility, different alternative living arrangements, and voluntary childlessness, have led some researchers to suggest that the traditional family is about to vanish (Cooper, 1986). However, evidence from a variety of studies (e.g. Kiernan, 2002) reveals that most people still establish a permanent relationship and continue to have children. Although the popularity of marriage is decreasing, studies have shown that cohabiting couples resemble married couples in many ways (Brown/Booth, 1996) and are able to fulfil one of the main functions of the family – socialisation of children.

An increase in divorce rates in the majority of western countries may also be interpreted as a threat to the existence of the traditional family. However, research shows (Kalmijn, 2007; Kiernan, 2002) that most divorced people are ready to establish new and enduring relationships. Some concerns about the future of the family are also related to low fertility levels. However, some fertility surveys have revealed that in the absence of extreme environmental pressures against having children, the vast majority of married and cohabiting couples want to have at least one child.

Thus, the thesis of 'the death of the family' based on a quick reading of population statistics is most probably not the most likely scenario for the future of the family. Beck has considered family a "zombie category – dead but still alive" (Beck-Gernsheim, 2002).

The second scenario – back to the traditional family?

This scenario speaks of restoration of the traditional family, but it is unclear what is meant by 'traditional family'. If it is the traditional economic family with the father as the sole breadwinner and the mother as homemaker, then in twenty-first century Europe with highly educated women and changing attitudes towards gender roles, it seems quite unrealistic to return to a family model with a stay-at-home mother. Alternatively, does the traditional family mean a family that excludes premarital sex and non-marital cohabitation? At least in Europe both are widespread, and it seems to be unrealistic to expect present sexual 'permissiveness' to be replaced by the strict sexual norms attributed to the first half of the twentieth century.

The third scenario – further increases in variation of family forms

Family sociology has demonstrated increasing variation in household types and more complex family life courses in recent decades (Jallinoja, 1994). Modernisation has led to acceptance of a variety of family forms based on the individual choice (Hoffmann-Nowotny, 1987; Beck-Gernsheim, 2002). It gives us reason to expect that the diversity of family types, including families based on same-sex relationships, will continue to increase in the future.

Increased geographical mobility and career opportunities offer the possibility or even facilitate the increase of a LAT relationship (living apart together). LAT relationships may be entered into for a variety of reasons or circumstances (employment location, mobility requirements, family phase, financial position, etc.). Due to psychological burdens and also economic problems, it is likely that LAT relationships will remain a minority among partnership choices (Trost, 1998).

In conclusion, we can say that the changes taking place in European family behaviour and in the family in the western world in general are extremely interesting and sometimes even unexpected, requiring further exploration and analysis. The future of the family is one of the issues that will provoke serious discussions among the researchers involved in FAMILYPLATFORM.

References

- Beck-Gernsheim, E. (2002). *Reinventing the Family: in search of new lifestyles.* Cambridge: Polity Press.
- Brown, S.L. & Booth, A. (1996). *Cohabitation versus Marriage: A Comparison of Relationship Quality.* Journal of Marriage and the Family 58: 668-678.
- Cherlin, A.J., Scabini, E. & Rossi, G. (1997). *Still in the Nest: Delayed Home Leaving in Europe and the United States.* Journal of Family Issues 18.6: 572-575.
- Cliquet, R. & Avramov, D. (1998). *The Future of the Family: A Sociobiological Approach.* In Matthijs, K. (ed.) *The Family. Contemporary Perspectives and Challenges.* Festschrift in Honor of Wilfried Dumon. Leuven: Leuven University Press, 159-180.
- Cooper, D. (1986). *The Death of the Family.* Harmondsworth: Penguin.
- Corijn, M. & Klijzing, E. (2001). *Transitions to Adulthood in Europe.* European Studies of Population, Vol. 10. Dordrecht, Boston, London: Kluwer Academic Publishers.
- *Europe in Figures – Eurostat yearbook 2009.* (2010). Luxembourg: Office for Official Publications of the European Communities.
- Hoem, J.M., Kostova, D., Jasilioniene, A., Muresan, C. (2007). *Traces of the Second Demographic Transition in four selected countries in Central and Eastern Europe: Union formation as a demographic manifestation.* MPIDR WORKING PAPER WP 2007-026. Rostock: Max Planck Institute for Demographic Research.
- Hoffmann-Nowotny, H.J. (1987). *The Future of the Family.* In *European Population Conference 1987, Plenaries.* Helsinki: Central Statistical Office of Finland, 113-200.
- Jallinoja, R. (1994). *Alternative family patterns; their lot in family sociology*

- *and in the life-worlds of ordinary people.* Innovation 7.1: 15-27.
- Kalmijn, M. (2007). *Explaining cross-national differences in marriage, cohabitation, and divorce in Europe,* 1990-2000. Population Studies 61.3: 243-263.
- Kasearu, K. (2007). *The case of unmarried cohabitation in Western and Eastern Europe.* Paper presented to the conference of European Network on Divorce "Comparative and Gendered Perspectives on Family Structure", 17-18 September 2007, London School of Economics.
- Kiernan, K. (2002). *Cohabitation in Western Europe. Trends, Issues and Implications.* In Booth, A. & Crouter, A.C. (eds.) *Just Living Together. Implication of Cohabitation on Families, Children, and Social Policy.* Lawrence Erlbaum Associates, Inc.: 3-31.
- Lanciery, G. (2008). *Population in Europe 2007: first results.* Eurostat: *Statistics in focus* 81/2008.
- *Mobility in Europe: Analysis of the 2005 Eurobarometer Survey on Geographical and Labour Market Mobility.* (2006). European Foundation for the Improvement of Living and Working Conditions. Available from: www.eurofound.europa.eu.
- *OECD Family Database.* (2008). *Fertility rates (SF 4).* OECD Social Policy Division, Directorate of Employment and Social Affairs: www.oecd.org/social/family/database.
- *OECD Family Database.* (2008). *Mean age of mother at first childbirth (SF 5).* OECD Social Policy Division, Directorate of Employment and Social Affairs: www.oecd.org/social/family/database.
- *OECD Family Database.* (2008). *Share of births outside marriage (SF 6).* OECD Social Policy Division, Directorate of Employment and Social Affairs: www.oecd.org/social/family/database.
- *OECD Family Database.* (2008). *Marriage and divorce rates (SF 8).* OECD Social Policy Division, Directorate of Employment and Social Affairs: www.oecd.org/social/family/database.
- *OECD Family Database.* (2008). *Cohabitation rates and prevalence of other forms of partnership* (SF 9). OECD Social Policy Division, Directorate of Employment and Social Affairs: www.oecd.org/social/family/database.
- *Population Statistics,* 2006 edition. (2006). Luxembourg: Office for Official Publications of the European Communities.
- *Population Statistics,* 2008 edition. (2008). Luxembourg: Office for Official Publications of the European Communities.
- Pryor, J. & Rodgers, B. (2001). *Children in Changing Families.* Life after Parental Separation. Oxford: Blackwell.
- *Recent Demographic Developments in Europe 2001.* (2002). Strasbourg: Council of Europe Press.

- Speder, Z. (2002). *Diversity of family structure in Europe.* Budapest: Demografia.
- *The Social Situation in the European Union 2009.* (2010). Luxembourg: Office for Official Publications of the European Communities.
- Trost, J. (1998). *LAT Relationships Now and in the Future.* In Matthijs, K. (ed.) *The Family. Contemporary Perspectives and Challenges.* Festschrift in Honor of Wilfried Dumon. Leuven: Leuven University Press, 209-220.

1.3 Changes in Finnish Families: Towards Full-Time Motherhood and a New Familialism?

Marjo Kuronen, Teppo Kröger and Kimmo Jokinen
Family Research Centre, University of Jyväskylä

Finland has become known internationally as one of the Nordic woman-friendly welfare states where extensive public day-care provision for young children has given women an "exit out of family responsibilities", thereby offering possibilities to combine family life and paid work. This interpretation has probably always been too ideal. Raija Julkunen (1992: 47) reminded us back in the early 1990s that the woman-friendliness of the Finnish welfare state needed to be critically analysed, because even the "best reforms for women in the whole world" might have unintended consequences.

Finnish women have been described as working mothers. Traditionally, Finnish women have worked full-time, including mothers with young children. What has also been typical for Finland is that the employment rate of lone mothers has been even higher than for married or cohabiting mothers: in the mid-1980s, 90% of all lone mothers were in paid work. The dual-earner family model was also strongest at that time in Finland (Haataja, 2004).

However, the situation has changed quite dramatically during the last 20 years. Full-time motherhood has become more popular and youngest children are taken care of at home, mostly by their mothers. Even if family policy actively encourages men's involvement in parenting, it is only very slowly changing gendered practices of child care. In the early 2000s, the maternal employment rate of mothers of children under the age of three was unambiguously low in Finland (32%) compared to the European, not to mention even the Nordic level (Lister *et al.*, 2007: 126). In this respect, Finland is moving in the opposite direction to most of the other countries of Europe. The reasons for this are a complex mixture of political decisions, changes in the economic situation and in working life, gender relations, and ideological changes in society.

In Finland, ever since the 1980s, there have been two simultaneous but contradictory trends in child care policy: gradual expansion of public day-care provision, and financial support for parental child care. This has very much been a political compromise. In 2005, the take-up rate of publicly financed day-care for children was much lower in Finland than in the other Nordic countries. For children aged 1-2 years old, it was 37% in Finland compared with 54-85% in the other Nordic countries, which can be explained by the extensive use of home care allowance schemes. Even in the older age group (3-5 years), the rate is clearly lower in Finland, at 69% and 91-95%

respectively (Eydal/Kröger, 2010: 25). However, it should be mentioned that for Finnish women full-time motherhood represents a temporary phase in life. The vast majority of mothers return to paid employment at the latest when their youngest child turns three, that is, when their eligibility for child home care allowance ends. Financial support for home care is a controversial issue: on the one hand, it is an important right for families with young children and official recognition for unpaid care work; on the other it maintains gender inequality and weakens the position of women in the labour market (Repo, 2009).

Financial support for home care is not the only factor in explaining the rapid change: a deep economic recession in the early 1990s strongly influenced the employment rate and had lasting consequences for the Finnish labour market. Its influence has been remarkable especially among lone mothers, whose poverty risk has dramatically increased since the early 1990s mainly because of growing unemployment but also because of cuts in family benefits. For some of them full-time motherhood has offered a more positive identity than unemployment even if it has also meant a financial struggle in daily life (Krok, 2009). There are also full-time mothers to whom it has offered an option to exit working life, which is increasingly characterised by a hardening of demands and a worsening of conditions. Since the 1990s, career options for even academically trained women have been increasingly insecure and fragmented.

Attitudinal changes have also taken place. Many feminist researchers talk about a new familialism, where families instead of the state - and especially women in families - are expected to take more responsibility for caring (Mahon, 2002: 150-3). Even a turn towards a new kind of full-time mother society (Anttonen, 2003: 178-9) can be recognised, where the rhetoric of "the best interest of the child" and "parental choice" has made the general attitudes towards paid work of mothers with young children more negative than before. This is rather new and unique phenomenon in Finnish society. This example from Finland shows not only that welfare state models are not everlasting, but also that there can be unintended backlashes against them.

References

- Anttonen, A. (2003). *Lastenhoidon kaksi maailmaa [Two worlds of child care]*. In Forsberg H. & Nätkin R. (eds.) *Perhe murroksessa*. Helsinki: Gaudeamus, 159-185.
- Eydal, G. B. & Kröger, T. (2010). *Nordic family policies: constructing contexts for social work with families*. In Forsberg H. & Kröger T. (eds.) *Social Work and Child Welfare Politics. Through Nordic lenses*. Bristol: Policy Press, 11-27.
- Haataja, A. (2004). *Yhden tai kahden ansaitsijan malli: vaikutukset ansiotyön, hoivan ja tulojen jakoon [Single or dual-earner model: influences on division of paid work, caring and income]*. In Hjerppe R. & Räisänen H. (eds.) *Hyvinvointi ja työmarkkinoiden eriytyminen*. Helsinki: VATT, 162-186.
- Julkunen, R. (1992). *Hyvinvointivaltio käännekohdassa [Welfare state in its turning point]*. Tampere: Vastapaino.
- Krok, S. (2009). *Hyviä äitejä ja arjen pärjääjiä - yksinhuoltajia marginaalissa [Good mothers and everyday survivors – Single mothers on the margin]*. Acta Universitatis Tamperensis 1437. Tampere: Tampere University Press.
- Lister, R., Williams F., Anttonen, A., Bussemaker, J., Gerhard U., Johansson, S., Heinen, J., Leira, A., Siim, B. & Constanza, T., with Gavanas, A. (2007). *Gendering Citizenship in Western Europe: New challenges for citizenship research in a cross-national context*. Bristol: Policy Press.
- Mahon, R. (2002). *Child Care: Toward What Kind of "Social Europe"?* Social Politics: International Studies in Gender, State & Society 9.3: 343-379.
- Repo, K. (2009). *Pienten lasten kotihoito – Puolesta ja vastaan [Home care of young children – For and against]*. In Anttonen A., Valokivi H. and Zechner M. (eds.) *Hoiva. Tutkimus, politiikka ja arki*. Tampere: Vastapaino, 219-237.

1.4 Short Account of Changes in the Family in Italy

Carmen Leccardi and Miriam Perego
Department of Sociology and Social Research, University of Milan-Bicocca

In Italy, as in the majority of western countries, the family has undergone considerable change and taken on a range of new forms, especially since the 1960s. In fact, in the course of the last thirty years it has become more and more common to see a variety of new family models – from the unmarried couple, to the single-parent family, to the reconstituted or recomposed family, to the "mixed" family (the family or couple made up of an Italian and a foreigner).

Quite apart from the degree to which each of these family forms has spread, we should emphasise the strong likelihood that each of us in the course of our lives will come into contact with this new multiplicity of forms of family life. Thus, analysing the principle forms of transformation that the family is going through means getting to grips directly with the actual life experiences of the individual – the young people involved, both male and female.

Official ISTAT statistics confirm that Italy, like other countries, is characterised by a progressive increase in the presence of unmarried couples or free unions. Moreover, although the overall number of such unions is still relatively low compared with the figures for the rest of Europe (in 2006, 637,000 out of a total of 14 million couples), the rate of growth has increased from the 1990s onwards – in the period between 1994 and 2006, from 1.6% to 4.5% of all couples (ISTAT, 2005). Notwithstanding this, for the majority of unmarried couples in Italy today living together still constitutes a kind of "preparation" for marriage and is regarded as a transitory phase of life (Sabbadini, 1997; Buzzi *et al.*, 2008). Compared with the rest of Europe, Italy continues to show a clear preference for the model of the traditional family (marriage with children generated within it) even though, as stated above, gradually increasing numbers of people (in particular among the younger generations) are choosing to live together.

It is interesting to focus attention in particular on how young people, both male and female, are postponing their exit from their family of origin and subsequent formation of a new family. Indeed, of the 27 European countries, Italy (together with other Mediterranean countries) is the country in which children wait the longest to leave their families of origin. In 2003 for example, the number of unmarried young people between 18 and 34 - male and female - who lived with at least one of their parents numbered 7,600,000 (60.2% of the total). The tendency to continue to live in the parental home even after having achieved economic independence has been called the "long family"[1], and it is also accompanied by new ways of creating families on the part of young men and women.

At a general level, it must be stated that in Italy the incidence of marriage has also fallen noticeably. In 2006 for example, the ratio of the number of marriages to the number of residents was 4.1 marriages for every 1,000 inhabitants, with a contraction between 2001 and 2006 of 7.8%. The birth rate in our country is also particularly low, 1.17%, while the European mean is about 1.5%.

The nature of relations within the couple, and their behaviour, are also changing, above all as a consequence of the emergence of a more equal relationship between the genders. This, at least within the private sphere, is contributing to making women more independent, furnishing them with a greater degree of liberty and decision-making power. In fact, in the case of unions that are no longer considered satisfactory, women often decide to cut the tie rather than maintain the relationship. Nevertheless, the frequency of divorce in Italy is still very low, lower not only than in northern European countries (50-55 per 100), but also if compared to other southern European countries (9 divorces per 100 in Italy, 18 per 100 in Greece).

Italy's territorial divide also finds expression in terms of marital breakdown. While the regions of central and northern Italy are closer to European norms - hence characterised by a more marked tendency to resort to separation and divorce, those of southern Italy are more traditionalist. In fact, southern couples are on average less inclined to resort to a definitive dissolution of the marital tie, even though the evidence shows that over recent years in this region too there has been an increase in the incidence of divorce (De Sandre *et al.*, 1999). In any case, the increased tendency of couples to decide to separate and divorce is producing forms of the family in Italy that are more articulated and complex than was the case a few decades ago.

Another phenomenon that it is worth focusing attention on is that of single-parent families. Although the number of such families in Italy - in contrast to other European countries - is relatively low, there has nonetheless been a significant increase. In particular, there has been a noticeable shift from the old form of the single-parent family, which was the outcome of unavoidable or unsolicited events (the death of a spouse, abandonment, etc.), to a new form of single parenthood: the result of a deliberate choice (Barbagli *et al.*, 2003). Figures for the years 2006-2007, for example, indicate that in Italy single-parent families - in line with the rest of Mediterranean Europe - constituted 5-10% of the total number of nuclear families.

---

[1] In Italy the expression refers to the prolonged presence of parents and children under the same roof.

In Belgium, France and Holland, by contrast, the incidence of single-parent families is greater: between 11 and 15%, while in northern Europe, in particular Great Britain and Germany, single-parent families make up a fifth of the total (Zanatta, 2008).

Finally, there is yet another type of family that is becoming more common in Italy just as elsewhere: the so-called "mixed family". In a period of ten years or so the number of marriages between Italians and foreigners has quadrupled, rising from 58,000 in 1991 to 200,000 in 2005 (ISTAT, 2004). This has been also accompanied by a 22% increase in the number of children born into mixed couples (ISTAT, 2004).

References

- Barbagli, M., Castiglioni, M. & Della Zuanna, G. (2003). *Fare famiglia in Italia. Un secolo di cambiamenti [Forming a Family. A Century of Changes]*. Bologna: il Mulino.
- Buzzi, C., Cavalli, A. & De Lillo, A. (2007). *Rapporto giovani. Sesta Indagine IARD sulla condizione giovanile in Italia [Report on Youth. The Sixth Enquiry of the IARD Institute into the Condition of Young People in Italy]*. Bologna: il Mulino.
- De Sandre, P., Pinnelli, A. & Santini A. (1999). *Nuzialità e fecondità in trasformazione: percorsi e fattori del cambiamento [Marriage and Fertility in Trasformation: Pathways and Factors of Change]*. Bologna: il Mulino.
- ISTAT 2005. *Population Statistics of ISTAT,* The Italian National Institute of Statistics.
- ISTAT 2004. *The Annual Report of ISTAT,* The Italian National Institute of Statistics for 2004.
- Sabbadini, E. *Le convivenze "more uxorio" [Cohabitations]* in M. Barbagli and C. Saraceno. (1997). *Lo stato delle famiglie in Italia [The Condition of Families in Italy]*. Bologna: il Mulino.
- Zanatta, A. L. (2008). *Le nuove famiglie [The New Families]*. Bologna: il Mulino.

1.5 Trends in the German Family Model: Pluralisation of Living Arrangements, and Decrease in the Middle-Class Nuclear Family

Ursula Adam, Loreen Beier, Dirk Hofaecker, Elisa Marchese, Marina Rupp
State Institute for Family Research, University of Bamberg

As in a number of other European countries, family arrangements in Germany in recent decades have become more diverse. Especially the incidence of the 'middle-class nuclear family' (a household with married parents and the biological children of both spouses) has decreased. This trend is driven by developments common to a lot of western countries: an overall decrease in and postponement of marriages, more divorces, low and delayed fertility, and a rising number of children born outside marriage.

A decline in marital unions, which are being entered into at ever higher ages, has led to an increase of the mean first marriage age, which has shifted from 25.6 in 1970[1] to 33 years in 2008 for men, and from 23 to 30 years for women. In the new *Länder* (the former German Democratic Republic) the first marriage age in 2000 was somewhat lower; however it increased faster than in the old *Länder*[2] since 1990, which points to a convergence between East and West Germany. This trend is complemented by an increasing divorce rate: while in 1970 there were 0.51 divorces for every 100 marriages, the number of divorces more than doubled by 2007 (in 2007 it stood at 1.03). However, the incidence of divorce is lower in the new[3] than in the old *Länder*[4] (0.84 vs. 1.07 in 2007). (*Statistisches Bundesamt*, 2009c)

Following the trend for delaying marriage, family formation shifted likewise: the mean age of married mothers at first childbirth[5] has increased by 5.8 years since 1970, up to 30.1 years in 2008 *(ibid.)*. At the same time the share of births outside marriage has risen significantly, from 5.5% in 1970 to 32% in 2008. Especially in the new *Länder*[6] the share of births outside marriage in 2008 was more than double the share of that in the old *Länder* (57.8% vs. 25.8%). Taken together with the trends in divorce, these trends point to a generally lower sig-

---

[1] Data for Germany in 1970 (territory of the Federal Republic of Germany before 1990).
[2] Excluding the Federal State of Berlin.
[3] Excluding East Berlin.
[4] Including East Berlin.
[5] Up until 2008 the German "Mikrozensus" (annually population census based on a sample of 1 per cent of German households) only collected data on the age of mothers at the first birth of a child born in a marriage.
[6] Including the State of Berlin.

nificance of the institution of marriage in the eastern part of Germany.
Following the delay in family formation, total (period) fertility rates (average number of children of all women between the age of 15 and 49) in Germany also show a long-term decrease: while in 1970 it was 2.02 (Heß-Meining/ Tölke, 2005: 231), by 1990 it fell significantly, to 1.45, with a less dramatic reduction by 2007 (to 1.37). The development in the new *Länder*, however, is unique in that fertility decreased sharply following the unification with West Germany from 1.5 (1991) to a record low value of 0.8 in 1994, but it has been gradually converging to the level of the old *Länder* since then (*Statistisches Bundesamt*, 2009c).

These trends are accompanied by notable changes in household structure, especially a decreasing number of people living in family households (with single children of any age in household). The share of those living in family households relative the whole population in Germany dropped from 67.2% in 1970 to 50.9% in 2008 (*Statistisches Bundesamt*, 2009b). The composition of family households also changed dramatically: in particular, the incidence of larger families with three or more children has declined (1970[7] 22% of families, in 2008 11.8%; Statistisches Bundesamt, 2010). Furthermore, there was a striking change in the importance of living arrangements: whilst in 1996 a distinct majority (84.4%) of children under the age of 18 were living in one household with their married parents, and only 4% lived in a cohabitation (couples, living together without being married) or same-sex household and 12% in a single parent household, these numbers had changed to 77%, 7% and 16% (respectively) by 2008 (Statistisches Bundesamt, 2009b, ifb-calculations).

As a result, there has been a decrease in the so called 'middle-class nuclear family': in 1996 about 81% of family households consisted of married parents with their minor children; however, by 2008 this figure had dropped to 73%. Accordingly the share of single parent households[8] increased from about 14% to 19% during the same period of time. Taking a longer-term perspective, the share of divorced and separated as well as of single parents has increased sharply while the share of widowed persons has correspondingly declined (since 1970[9]). An increasing number of these single parent households[10] are composed of mainly sole-parent mothers (87%) (*ibid.*, ifb-calculations).

Alongside the decrease in family households with married parents with their minor children and the increase in single parent households, Germany has seen a moderate increase in cohabiting parents, from 5% of family

---

[7] Findings of the 1970 population census.
[8] With minor children without age limit.
[9] Findings of the 1970 population census.
[10] Including cohabiting parents until 2002.

households with minor children in 1996 to 8% in 2008. This share of cohabiting parents is about three times higher in the new *Länder*[11] than in the old *Länder (ibid.)*.

Finally, according to data from the Generation and Gender Survey, in 2005 about 14% of family households in Germany were stepfamilies with minor children: 9% families with children of one partner and 5% with at least two children not directly related (Steinbach, 2009: 165f). The majority (about 69%) were stepfamilies with stepfathers, a third (about 27%) stepfamilies with stepmothers and a small minority (about 4%) with stepfathers and mothers (GGS, 2005; Steinbach, 2009: 167). Overall, stepfamilies are more common in the new than in the old *Länder (ibid.)*.

The described trend towards a pluralisation of living arrangements and the simultaneous decrease in the middle-class nuclear family in Germany resembles the experience of most European societies. However the different development of family models between the new and the old German *Länder* appears to be unique in Europe: the end of the "Golden Age of Marriage" with its high fertility rates, an almost completely married generation, low divorce rates and an early start to the building up a family, has in the old *Länder* been accompanied by the pluralisation of family forms since the 1960s. In contrast, in the new *Länder* the model of the "nuclear family" was more stable until the 1970s. However, dramatic changes in family factors in the new *Länder*, reflected in the sharp decline of fertility and marriage and a steep rise in divorce rates, were almost simultaneous with the reunification of East and West Germany. One reason was the ending of political support for the model of the "nuclear family", as well as rising insecurity throughout the political transformation. Since the middle of the 1990s, however, the situation has become more stable, and fertility as well as marriage and divorce rates have started to increase again, though they have not reached the same level as in the old *Länder* (see Peukert, 2008: 341ff).

References

- Heß-Meining, U. & Tölke, A. (2005). 4. *Familien - und Lebensformen von Frauen und Männern*. In Cornelißen, W. (ed.) *Gender Datenreport. 1. Datenreport zur Gleichstellung von Frauen und Männern in der Bundesrepublik Deutschland*, 224-277.
- Peukert, R. (2008). *Familienformen im sozialen Wandel*. Wiesbaden: VS Verlag für Sozialwissenschaften.
- Statistisches Bundesamt. (2009a). *Bevölkerung: Eheschließungen, Geborene*

---

[11] Includes the State of Berlin.

- *und Gestorbene* 2008 nach Kreisen, Wiesbaden.
- Statistisches Bundesamt. (2009b). *Bevölkerung und Erwerbstätigkeit: Haushalte und Familien. Ergebnisse des Mikrozensus 2008,* Fachserie 1, Reihe 3, Wiesbaden.
- Statistisches Bundesamt. (2009c). *Bevölkerung und Erwerbstätigkeit: Natürliche Bevölkerungsbewegung 2007,* Fachserie 1, Reihe 1.1, Wiesbaden.
- Statistisches Bundesamt. (2010). Available from: *https://www-genesis. destatis.de/genesis/online;jsessionid=C5CAAFA1EA1481601922DC94E5DE9 75F.tcggen1.*
- Steinbach, A. (2008). *Stieffamilien in Deutschland. Ergebnisse des "Generation and Gender Survey" 2005,* in Zeitschrift für Bevölkerungswissenschaft 33.2: 153-180.

1.6 Families in Hungary

Zsuzsa Blaskó
Demographic Research Institute, Hungary

Although the timing, pace, and characteristics of family trends are national in nature in all countries, the main trends seen in families in Hungary are similar to those seen in other European countries. These include an ageing society, an increasing popularity of cohabitation without marriage, a decreasing number of births, together with a growing proportion of births out of wedlock, as well as increasing fragility of relationships.

Two-thirds of Hungarian private households were a family-type household in 2005. More than half of them were based on a married or cohabiting couple with or without children; the rest of the families (10% of all the households) consisted of a single parent with one child or more. Sole-person households accounted for almost 30% of all the households. Multiple families or multiple generations living together were rare (Hungarian Central Office).

Typically, there are one or at most two children in a Hungarian family today. In more than 50% of the families with children there is only one child, and there are two in another 35%. At the same time 10% of families are a "large family" with more than three children. Comparing the 2005 ratios to data from previous years shows a slow decrease in the percentage of couple-based households together with an increasing number of sole-person households. These tendencies are partly attributable to an ageing of the population, which is at least partially associated in turn with a growing number of widowed persons living on their own – in most of these cases these widowed people are women. Single persons at younger ages are also a significant and slowly growing subgroup of those living by themselves, although it is not usually an affordable option for young people in Hungary.

With just over 10% of households consisting of a sole-parent family, Hungary was at about the middle of the European ranking list in 2001, in between Spain (9.9%) and Ireland (11.7%). In Hungary – just like in most of the European countries, sole-parent families are typically headed by the mother rather than the father.

The proportion of sole-parent households has remained relatively stable since 1990 – though the divorce rate has increased and the proportion of births to unmarried women has also increased. At the same time, however, there have been more childless marriages among those breaking up in 2005 than in 1990, and in 2005 the majority (two-thirds) of births to unmarried women took place in a cohabiting partnership – rather than to a single mother.

Indeed, cohabitation has spread rapidly as a form of living in the last two decades. In 1990 only 4% of families were based on cohabitation, but this number had tripled by 2005. Not only has the number been changing: while previously it was very often the widowed or divorced elderly who chose not to marry, today this form of family formation is much more popular among younger generations. In fact, for many people today cohabitation is a transitory form of living that will either be turned into marriage after a test period or ended. Nevertheless, seven out of ten first cohabiting relationships starting between 2000 and 2004 took the form of cohabitation without marriage. This tendency took place alongside increased social acceptance of cohabitation, although this form of partnership is still considered a kind of test period prior to marriage itself, which would ideally follow the success of the test period. Statistics from the Hungarian Central Office show that cohabiting partners remain childless more often than married couples, however, and they are also more likely to have one child rather than more.

Chapter 2: Solidarities in Contemporary Families

Editorial

Carmen Leccardi and Miriam Perego
Department of Sociology and Social Research, University of Milan-Bicocca

This chapter of the *FAMILYPLATFORM Journal* is dedicated to a theme that is of great importance not just for families but for contemporary societies as a whole: intergenerational solidarity. Paradoxically, it seems, the growing tendency towards individualisation has been accompanied in more recent times by the rediscovery of forms of solidarity, at times quite unprecedented, within the family. Today, the various generations that make up the family - ever more frequently, as a consequence of demographic changes, consisting of as many as four generations - seem to be engaged not so much in conflict as in a continuous contest to offer solidarity.

The traditional conflict between older and younger generations, characteristic of western societies in the twentieth century, exploded, as is well known, with particular virulence in the sixties and seventies, the years of the youth and political movements. Starting from the nineties, thanks in large part to the spread of ever less authoritarian family relations (as Claudine Attias-Donfut underlines in this chapter of the journal), forms of comprehension, help and reciprocal support between the various components of the family have been rediscovered as a major resource in the resolution of problems confronting the various generations in social life.

Simultaneously, the turn of the new century has seen the emergence and spread of new expectations of family solidarity. These involve in an analogous way both the young and the less young. The young, for example, confronted by 'precarity' and instability in the labour market, expect to receive economic and relational support from their family; young adults expect help in fulfilling their new parental responsibilities; and the elderly expect support in confronting the material, health and psychological difficulties that advancing age brings with it. And in fact - this needs to be underlined - all these expectations are brought to bear on the baby-boomer generation. Today's fifty/sixty-year-olds thus find themselves at the centre of converging expectations. It is no accident that the French scholar Claude Martin has recently defined this generation as the 'pivot generation': a generation destined to carry on its shoulders multiple generational pressures, often difficult to reconcile.

It is in fact the first time in the history of humanity that such a large number of generations find themselves living together in the same historical time and on the same social scene. A situation, as is highlighted by the articles contained in this edition of the journal, capable of generating a scenario that was unthinkable up to a few decades ago – a scenario full of positive features but also, inevitably, of contradictions and ambivalences. Indeed, this latter characteristic constitutes a central theme of the interview with Ariela Lowenstein.

In this scenario an unprecedentedly central role is played by grandparents. The growth in the period of life in which people are grandfathers and grandmothers in good health and active on the social scene - albeit no longer in the labour market - is in fact continual. This new reality has changed not only the social profile of grandfathers and grandmothers and representations of them, but also the role they are able to play in offering active support to other members of the family: no longer just care-receivers, then, but also care-givers. Consider, for example, as confirmed by European data (taken into consideration here in particular by Francesco Belletti), the caring capabilities that grandparents demonstrate in respect of grandchildren, especially those not yet of school age – a form of help that, in contrast to others, is particularly widespread in southern Europe, where the welfare system is less extensive. Although in this respect too the gender variable is of crucial importance (grandfathers and grandmothers do not furnish the same amount, or quality, of care time: this theme is taken up in the interview with Carla Facchini and Marita Rampazi), there can nonetheless be no doubt as to the positive role that both exercise in the vitalisation of forms of solidarity within the family: both through financial and non-financial help.

In short, it is necessary to reconsider the prevalent notion that the elder generations are the exclusive recipients of help provided by the younger generations. It is also appropriate to distinguish, as is also underlined in other articles in this edition of the journal, between the elderly and the 'old elderly'. It is above all the care of the latter that has constituted in the last few decades a problem of great strategic importance in the increasingly older societies of Europe. There is no doubt that this situation is exacerbated by the growing instability of the family, together with the fact that an increasingly large number of adult women cannot undertake unpaid labour within the family, on account of their involvement in the labour market. Nevertheless, it would be an error not to draw attention to the other side of the coin: the 'young' grandparents that distinguish themselves by their capacity to play an irreplaceable role in the practice of forms of family solidarity.

It is important to remember, however, in relation to the question of family solidarity, that support and reciprocal help that continues to originate from the family is not and cannot be considered to be a substitute for public support (as Attias-Donfut rightly underlines). In fact, whatever the form and degree of support of public policies, and whatever their actual capacity to respond to needs, solidarity within the family tends to combine with public services rather than replace them. The variety of forms and manifestations of solidarity can therefore be explained at least in part by starting out from the differences in welfare policies in national and regional terms. Account must always be taken, albeit in terms of the variety of situations in question (for example, in respect of so-called 'large families': in this regard see the thoughts of Raul Sanchez) of the indisputable strategic importance of solidarity between the generations in guaranteeing the wellbeing of the family.

In the final analysis, public and private can come together constructively to confront problems - and overcome the social obstacles and uncertainties characteristic of our times - that fall on the shoulders of families. Nonetheless, a certain number of more general goals - for example, promoting dialogue and awareness between generations and actively involving the elder generations in solving the problems that relate to them - remain the specific responsibility of the public sector (as is documented here in the article by Lorenza Rebuzzini, who illustrates the outcomes of initiatives undertaken by Turin and Manchester municipal councils to this end).

In conclusion, the various points of view expressed by scholars and exponents of the world of family associations in this chapter of the journal confirm our direct experience: today, solidarity between generations within the family appears more alive and vital than ever - and also more and more alive, we might add, the more the future becomes gloomy. At the same time, taken as a whole, these testimonies induce us to reflect on an important strategic feature of this reality, i.e. the increasingly social nature of this help and solidarity. These forms of help and solidarity thus emerge as outcomes of specific historical circumstances, which have generated requirements and needs of an unprecedented nature in terms of support between the generations.

2.1 How Social Change is Transforming Relations Between the Generations

Interview with Claudine Attias-Donfut
Caisse nationale d'Assurance vieillesse

❖ **What has brought about the biggest recent changes in intergenerational solidarities?**

Intergenerational solidarities have always existed, but they have become more prevalent in recent decades due to three main conditions:

- demographic changes related to lengthening life expectancy;
- changing values and attitudes that have profoundly affected the family;
- last but not least, the development of social protection systems embodied in the welfare state: because intergenerational solidarities complement and are even conditioned by public solidarities.

1. Demographic changes

Lengthening life expectancy impacts on all stages of life: youth is extended and old age is deferred in the sense that people are living longer in better health. People are also grandparents for longer – in some cases up to half their lifetimes.

This rise in life expectancy is producing so-called "vertical" multi-generational families (three, four, or even five) with very few members in each generation. This differs from families in traditional societies, which are more "horizontal" in the sense that they have more children but fewer generations coexisting. Many families now have more grandparents and great-grandparents than grandchildren. This is a significant reversal of family age structures.

2. Changing family relations

Increasing gender equality and a declining patriarchal system have also produced profound changes in intra-familial relations. Education has become less authoritarian; the generations have grown more self-reliant, starting with the oldest, which has seen its standard of living improve and is increasingly co-residing less with other generations. Parent-children 'co-residentiality' has also increased, as young people are spending longer in education and finding it harder to integrate into the labour market.

These changes have produced a diversification of family models: the middle-class family model (two parents with two children) still exists, but is only one of others like blended families, lone-parent families, and so on. These models have brought about a new kind of family mindset that seeks to balance interdependence between family members with personal autonomy: the "freedom in togetherness" described by sociologist François de Singly.

Clearly, these changing family relationships are also affecting the bonds connecting the generations, but another factor has been more crucial still: the development of social protection.

3. The development of social protection

The development of social protection has particularly benefited young people and pensioners. Financial support has enabled young people to continue their education, but not without changing their status: children and adolescents are viewed more as adults in the making than as producers, as they still were in mid-twentieth century. This support has also encouraged parents to focus on their children's education and strengthen intergenerational solidarities by supporting them in education.

Universal entitlement to - and higher - pensions has given more financial autonomy to the adult generations and reversed the direction of solidarity flows. Many currently elderly people started work at a very young age (in the fields or mines), handing over their entire pay packet to the family. Solidarity flowed from the children to the oldest family members. This has now reversed, as the development of social protection has given the older age groups financial independence so that they are no longer financially dependent on their own adult children.

❖ *What are the particular forms of non-monetary intergenerational solidarity today?*

There are at present several kinds of non-monetary support between generations:

- personal care for elderly people and children, people with disabilities, or adults who have care needs at some point in their lives;
- co-residentiality with one's parents, children or grandchildren;
- grand-parental childcare for grandchildren.

In addition, there are other forms of practical help:

- 'odd jobs' about the home, gardening, transport, domestic chores;
- administrative help such as form filling, tax returns, health and social security, etc.

Who gives to whom? Who receives from whom? These questions were answered by the findings of one of the first surveys done on intergenerational solidarities – a tri-generational study on a sample of nearly 5,000 people representing 2,000 three- (sometimes four-) generation families living across France but not necessarily in the same household. The representatives of each generation were asked what they had given to and received from the other two in the previous five years.

Unsurprisingly, the higher-educated, higher income groups were givers. However, the low- and middle-income givers gave more than the high-income groups proportionate to income: in other words, the lower earners did more.

Where does the giving go? Mainly to the children as support while in education, unemployed, unmarried or at risk of social exclusion. In these cases, the giving partakes of an investment in human capital. However, more is given to girls than boys, simply because girls tend to stay in education longer.

Who gives most? Chiefly, the pivotal generation - the family founders - but also the childless through giving to their nephews, nieces or collateral relations, etc. The truth of the old adage *a father gives more to ten children than ten children to their father* was also borne out by the findings.

The rate of giving decreases with ageing, but not in retirement: the amount of giving is the same before and after retirement, and even increases. Giving declines throughout old age, but never stops entirely. It decreases with the move into advanced age. Giving decreases for children but not for grandchildren (who are, by then, in higher education).

The survey[1] distinguishes between cash gifts and services:

- Cash gifts: the oldest generations were found to give most money to the so-called pivotal generation and the grandchildren; those in the pivotal generation give to their children, but little to their own parents; very few young people give to their parents (none give to their grandparents). In other words, financial solidarities flow

---

[1] The research is that mentioned above, i.e. tri-generational research on a sample of nearly 5,000 people representing 2,000 three- and four-generation families living across France but not necessarily in the same household.

downwards between the generations. This may seem self-evident because that is the experience of most of us, but historically it is a new phenomenon: before universal social protection came in, and even before it was improved, solidarity flowed upwards (children started working young to help their parents).
- Services (care, childcare, etc.): these are more evenly distributed between the pivotal generation (which provides help to dependent parents) and grandparents who provide occasional childcare.
- State help goes more towards the older (through old age benefits and pensions) and younger (through study grants) generations than to the pivotal generations (who have a greater income tax burden).

The methodology used to evaluate services in percentage terms was that the nearly 5,000 respondents interviewed – 1,958 from the pivotal generation (aged 49-53), one of their parents (1,217, average age 77) and one of their adult children (1,493, average age 27) - were asked to specify the frequency and time represented by each service provided (help with housework, loan of a car, care, help with homework, shopping, etc.). The results were then valued in cash terms. This showed that the family contribution (more often made by women than men) is a significant effort that justifies being described as a "domestic economy". These findings also highlighted the special role of the pivotal generation (a concept which has since come to be recognised) represented by the fifty-somethings.

One of the kinds of help provided deserves fresh attention: grandparental childcare. It has always been there, but it is now taking a new form. In the research, the sample groups were asked to specify what help they received from their parents. The finding is that young people today are receiving more help than the two earlier generations. Today's grandparents are spending more time with their grandchildren.

This might seem illogical given the increase in collective support through the expansion of nursery schools and day nurseries. The explanation for it lies in the increased needs of young working couples. Not only are women working, they are also engaging more with their careers. And young couples also want more "me time", and so often draft in both sets of grandparents. Meanwhile, grandparents have fewer grandchildren. But it is less common today for grandparents to have direct responsibility for bringing up their grandchildren – mainstream psychological opinion goes against it, stressing the importance of parents raising their children themselves. In contrast, grandparents are more readily enlisted for occasional help to look after a sick child, for example, or to supplement public childcare provision. This increased involvement by grandparents can also be put down to the higher rate of marital breakdowns. Grandparents are the first-line bulwark to cushion the effects of family crisis whose help is more readily enlisted in the event of divorce. In fact, the first US surveys on the role of grandparents were prompted by the grandparental role in relationship breakdowns.

Contrary to popular beliefs on family decline, family solidarities are ultimately more enduring today than ever before and more reliant on the grandparent generation. At the end of the day, what impact does grandparental support have? A study carried out by economists is informative: it seems to help young mothers get into and stay in the labour market. Their availability also influences the decision to have a career. Added to this is their contribution to child-raising and giving roots in a family tree. Through transmission, they act as "resident historians", to quote one interviewee.

❖ *How does France compare to other European countries in terms of intergenerational solidarities?*

A European comparative study done in 2004 and repeated in 2006[2] - SHARE (the Survey of Health, Ageing and Retirement in Europe) - shows commonalities and distinctive features.

The common trends included:

- enduring intergenerational solidarities, whatever the level of self-absorption and type of social protection;
- the pivotal generation plays a central role in every country (flows to the younger generations for monetary support, more generalised flows for time-based help);
- financial transfers flowed downwards through the generations in all welfare systems, while social support tended to flow upwards (with the exception of grandparental help).

But there are several differences, too:

- Sweden, Denmark and the Netherlands have a higher proportion of people involved in exchanges of practical support and care,

[2] The same surveys were carried out simultaneously in a dozen European countries plus Israel in 2004 and 2006 on the same individuals (longitudinal surveys) supervised by Axel Borsch-Supan (Mannheim Institute of Economics of Ageing) under the aegis of the European Commission. See *http://www.share-project.org/*.

- but these are inter-household, occasional flows. The mutual assistance network is largely composed of family members, but also of a substantial minority of non-relatives.
- In Spain, Italy and Greece, by contrast, exchanges are exclusively focused on family members, especially within the household. They are regular and intensive, and only occur between a handful of people.

The other countries - France, Germany, Austria and Switzerland - fall between these two groups, combining elements of tight local networks and more widely dispersed networks.

❖ *What is the relationship between public support and intergenerational solidarity?*

Contrary to another popular belief, the frequency of help is not greatest in the more family-centric, co-residential southern Europe, but rather in the northern countries, which also have the most extensive social solidarity networks that include both family and friends. But northern Europe is also where we find the most public support for children, the disabled and the elderly. So family support networks complement this, and this enables a larger number of people to be reached.

This European comparison between public and private solidarities is borne out by SHARE. It highlights the fact that those who were receiving public support in 2004 but not in 2006 did not benefit from increased family solidarity. Conversely, those who did not receive public support in 2004 but did in 2006 suffered no loss of family solidarity. In other words, an increase in public support is not a disincentive to family help. This is clearly visible when tracking the same families over time: where public support decreases, the family does not necessarily step in; the two forms of support complement and play into one another. It might even be said that the reason why the family has been able to maintain its role is because the development of social protection has brought order to the relations between generations precisely by giving the family the resources needed to provide support to one or other of its members. Most of the surveys show this. Some have implied a substitution effect between private and public help. But what we are actually seeing is a change in the nature of the help provided through intergenerational solidarity, such as where home help is brought in to do the work previously done by a family member. Family members will continue to help, but in other ways (keeping company, shopping, etc.). Likewise, study grants enable the family to do more to support young people in education. In this way, public support gives leverage.

A family does not live in a self-contained world, but in an environment on which it depends for its inputs. Without those environmental inputs, it may collapse. In a crisis, the family pulls together to help its members in difficulty, but if the crisis endures, and no public support is forthcoming, the family will eventually become depleted.

❖ *What is the link between intergenerational solidarities and inequalities?*

Where inequalities are concerned, the effects pull in opposite directions depending on which transfers we are looking at. There is no doubt that transmissions of assets are a factor - or even a perpetuating factor - in widening inequalities. In France, half of all inheritances include the transmission of a house which is often kept, or a gift which is used to buy a house. In both cases, the transmission further widens equality gaps.

Various studies confirm that in comparable social circumstances, two young couples will have different life courses depending on whether or not they have benefited from a transmission of assets: those who have been gifted money get onto the housing ladder earlier, with a smaller mortgage. And home ownership has knock-on effects throughout the rest of life.

In-life transfers, however, have the reverse effect: they benefit those most in need, where there are several children. This is because amongst siblings, the most successful highest earners tend to help out those who are having difficulties. Those in the most difficult circumstances receive the most from their parents. Similarly, it is elderly parents on the lowest incomes with children who earn more who receive help from their children.

In short, there is a tendency to balance living standards within the family, with the better-off paying for the worst-off. So solidarities operate to reduce intergenerational inequalities.

❖ *What are the consequences of the current crisis? And what are the future prospects?*

The first challenge is the problem of pension funding: we are already seeing public pensions falling and an increase in private insurance-based systems that only the best-off can afford. The big question mark is how the labour market will develop. A return to full employment would largely resolve the problem. But demographic trends - against the background of an ageing population - raise the spectre of shrinking resources to meet growing needs.

The second challenge is changing attitudes, where the trend is towards greater individualisation, especially among young people. This trend is reflected in a reduced willingness to make life sacrifices and, beyond that, a belief that the existence of public help and private services enables support to elderly parents to be outsourced. This trend illustrates the process of "denaturalisation" of help identified by Guberman/Lavoie (2008), in a Quebec study: what was seen as natural (i.e. taking care of others, for the older generations) no longer is. It is felt to be the community's responsibility to organise itself to provide help to those in need.

There is therefore a risk of a return to large pockets of poverty among pensioners, wider inequality between families, the risk of a polarisation between the casualties of the crisis who are bereft of family solidarities, and the unscathed who benefit from such solidarities. Pensioners will be less able to help their families and will need even more help from them.

This latter scenario requires social policies to be redefined either:

- by reorienting social protection to target the worst-off (as practiced by 'Anglo-Saxon tradition' countries) and letting the market meet the needs of the rest, in particular through the development of a private insurance system for elderly people with care needs while the state funds coverage for the poorest groups; or
- by pursuing a proactive policy to reduce income disparities: this is the path chosen by Nordic countries, where the entire population receives public support in return for reduced income disparities.

What makes social policy reforms even more important is that the harm done to the social protection system often cannot be undone.

❖ *Isn't the idea of intergenerational solidarities just a way of diverting public attention from falling social protection standards?*

It can be. Hence the need to stress how these solidarities and public solidarity play into one another. The family cannot take the place of public help because it is sustained by it. Strong family-centric attitudes as found in southern European countries are not enough to develop intergenerational solidarities. You still need a system of substantial public help. Families need this public support to continue doing what they do.

❖ **What can local authorities do to support that interaction?**

Local authorities can help through initiatives promoted by self-help groups and even cultural organisations: by helping to bring people from different generations together, they help build an attachment to society and have a decisive impact on family functioning.

Voluntary organisations act preventively if only by giving young people something to do, so that they are not left to their own devices. Unfortunately, active voluntary groups are the first casualties of cuts to subsidies in a recession.

❖ **You talked about the attachment to society created by private and public solidarities. But surely employment is the main pathway to inclusion?**

Employment is certainly a key issue, especially as the financing of social protection depends on it. But we must not undervalue the role of family solidarities in helping to find a job and even get into work. Families play a key role, both in helping to steer people towards good training courses, or leveraging their social networks to identify job vacancies.

Even someone in work needs intergenerational solidarity to organise their life, look after their children if any, and so on. In short, a job alone is not enough. You need to be in social networks. Support from the family can help to improve the way we live.

❖ **How do migration and intergenerational solidarity play into one another?**

The literature on the subject presents contrasting views of the relationship. One, following the modernisation theory, argues that migration speeds up or itself partakes of the modernisation process: this theory argues that acculturation brings about a more individualised lifestyle, a change in gender relations, a weakening of authority structures and a change in relations with the extended family. These structural changes interact with cultural changes to create sources of conflict between young people, whose adjustment to the host society culture is more rapid, and their parents who live in a form of biculturalism in which the cultures of the country of origin and the host country mingle without blending.

Contrasting with this picture of conflicted families on the brink of disintegration, the other view depicts immigrant families as actually typified by great cohesion, closeness and intergenerational solidarities, more acutely family-centric, responding to a need for protection in an alien or hostile environment.

Although apparently at odds, these two portrayals are not inconsistent: solidarity can exist alongside different cultural orientations, as well as with conflicts, which also embraces the ambivalent relationship which the insights offered by the sociologist Kurt Leuscher tell us characterises all generational relations. There is much to be said here for looking more deeply into the impact of these within-family relationships on the pathways to inclusion or integration of immigrants and successor generations. This would also include exploring their macro-social and intercultural implications, because intergenerational ties in transnational families are powerful vehicles of two-way cross-cultural influences between emigration and immigration societies.

These influences are particularly significant and fast-acting on the vexed issue of women's status. Significantly, more women than men choose to stay and integrate into the host country when it affords them greater freedom and equality than they had in the country of origin. And if they do return for any reason, they can become active agents of change to that effect. In other cases, the influence is exerted through those who have not emigrated but kept up lasting, long-distance ties with those who have.

Just to close off these few brief thoughts on migration, let me say a word about the interest of a read-across approach to the two big demographic trends, ageing and migration, which produce specific phenomena like retirement migration, i.e. return migration by those who want to spend their retirement years elsewhere than where they have lived and worked. Female labour migration is also expanding to meet the growing needs of the elderly services sector, which has far-reaching consequences for the intergenerational family ties of these migrant women. This is a big issue which, although not recent and already researched to some extent, seems destined to loom even larger.

Let me conclude by emphasising that while these few thoughts reflect my roots in French society, they apply to all societies - adjusted to suit the context, obviously - but especially that future research should wherever possible be both international and comparative in approach. The paradigm of intergenerational relations contains universal aspects that only international comparisons can bring to light. In the meantime, we must continue to explore the matter and go beyond the debates that tend to reduce the whole issue to a generation gap, which simplifies a far more complex reality.

References

- Attias-Donfut, C. (1988). *Sociologie des générations.* Paris: PUF.
- Attias-Donfut, C. (1991). *Générations et âges de la vie.* Paris: PUF.
- Attias-Donfut, C. (1995). *Les Solidarités entre générations.* Paris: Nathan.
- Attias-Donfut, C., Lapierre, N. (1997). *La famille Providence. Trois générations en Guadeloupe.* Paris: La documentation Française.
- Attias-Donfut, C., Segalen, M. (2007). *Grands-Parents. La famille à travers les générations.* Paris: Odile Jacob.
- Attias-Donfut, C., Arber, S. (eds.) (2007). *The Myth of generational conflict; Family and the State in an ageing Society.* London: Routledge.
- Attias-Donfut, C., Segalen, M. (2001). *Le siècle des Grands-Parents.* Paris: Autrement.
- Attias-Donfut, C., Lapierre, N., Segalen, M. (2002). *Le Nouvel Esprit de Famille.* Paris: Odile Jacob.
- Attias-Donfut, C. (2006). *L'Enracinement. Enquête sur le vieillissement des immigrés en France.* Paris: Armand Colin.
- Attias-Donfut, C., Wolff, F. C. (2009). *Le destin des enfants d'immigrés.* Un désenchaînement des générations. Paris: Stock.
- Guberman, N., Lavoie, J.P. (2008). *Babyboomers and the "denaturalisation" of care,* presentation at *International Workshop on Intergenerational Transfers and Support and their Linkages to Health and Well Being of Elders and Family Carers,* New Agendas For Research And Policy, University of Haifa, Israel, 29-31 October, 2008.
- Rapoport, R. N., Fogarty, M. P. & Rapoport, R. (1982). *Families in Britain.* London: Routledge & Kegan.
- Rothausen, T. (1999). *'Family' in organizational research: a review and comparison of definitions and measures.* Journal of Organizational Behavior 20: 817-830.
- Scanzoni, J. (2001). *From the Normal Family to Alternate Families to the Quest for Diversity With Interdependence.* Journal of Family Issues 22.6: 688-710.
- Skolnick, A. (1981). *The family and its discontent.* Society 18: 42-47.
- Stacey, J. (1996). *In the Name of the Family: Rethinking Family Values in the Postmodern Age.* New York: Basic Books.
- Strong, B. & DeVault, C. (1993). *Essentials of the Marriage and Family Experience.* St. Paul, MN: West Publishing Company.
- UNECE (1998). *Recommendations for the 2000 Censuses of Population and Housing in the ECE Region. No 49.* New York: United Nations.
- White, J. M. (1991). *Dynamics of Family Development: A Theoretical Perspective.* New York: The Guilford Press.

2.2 Family Solidarity and the New Forms of Social Uncertainty

Interview with Carla Facchini and Marita Rampazi
University of Milan-Bicocca / University of Pavia

❖ *Professor Facchini, in your view, who are the elderly today? How can they be defined?*

Before answering this question it's necessary to recall that in the course of the last few decades average life expectancy has increased dramatically, rising from about 65 years of age at the beginning of the fifties to about 70 in 1970/1971 and almost 80 in 2010. This means that, in contrast to the first half of the last century, after reaching 60-65 years of age - the age at which in the statistics the label "elderly" gets applied - the majority of people can expect to live on average another 15 to 20 years. Moreover, it is quite likely that these people will spend over half of this time in physical and economic circumstances not unlike those they enjoyed in their mature adulthood. So it is completely misguided to view the elderly as a homogenous block. Even more than with adults, the elderly are characterised by a wide variety of conditions both in terms of their health and self-sufficiency and in terms of their economic resources, family type and degree of social inclusion.

This multiplicity is without doubt related to gender, social conditions, the socio-economic characteristics of the geographical contexts in which their lives unfold and the various forms of welfare available in them. But just as fundamental is the role played by age - in all its aspects. By this I mean both age understood as a progression along the life course and age conceived in terms of subjects belonging to a particular generation. Marita Rampazi and I use this term in the sense that Mannheim attributes to it[1], in that we want to underline that the cohorts born between the first few decades of last century and the fifties experienced in their youth, or in other words, in the life phase that is most important from *the* point of view of biographical 'projectuality', historical events that had huge symbolic significance for the construction of their identity. These events took the form above all of the war and (especially in Italy and Germany) the transition to democracy for the older generations and of the political movements of the late sixties for the following generations. Moreover, in many European countries, particularly those like Italy, Spain or Portugal, which underwent modernisa-

---
[1] I refer here to Karl Mannheim's reflections on the generations. See Mannheim (1952).

tion (industrialisation, universal education, secularisation and the loss of the importance of traditional family membership) at a later stage, these 'historical' generations also took the form of fully fledged 'social' generations.

If we consider the "old elderly", or those who are older than 85 today, we should bear in mind that generally in Italy these people did not go to school beyond primary school, and began to work at a very young age as unskilled workers in industry, construction or agriculture. These are sectors that were characterised by physically very demanding work and a low level of skill, but which also witnessed the development of increasingly extensive social security provisions. The extreme precarity associated with the war was the thing that marked this generation more than anything else (or, more precisely, that constituted it as a generation in sociological terms). On the other hand, in cultural terms, this generation enjoyed firmly established certainties of another form, in a context still strongly influenced by pre-modern traditions, especially as far as the structure of the family and the nature of family roles were concerned. The ethic that characterised it was centred, for men and women respectively, on economic production and biological reproduction. In the course of their lives these people also came to know, or better still, acquired 'new' certainties, thanks to the development in the postwar period of systems of civil and social security. These new certainties were added to those already in existence, thereby contributing to the construction of a cultural model in which collective 'progress' was closely intertwined with individual advancement – obviously on condition that the beneficiaries respected the restrictions imposed by their given work and family roles.

Let us now turn to the following generation, the generation of people born between the mid-thirties and the war years. These subjects enjoyed at least a basic education and rarely started working before the age of fourteen. Generally, they worked in industry or the tertiary sector, and though they did jobs that were often physically demanding and of a limited skill level, they identified to a significant extent with their work – an identification which was underpinned by regular wage or salary increases and improvements in conditions and also by some form of career progression. In its youth this generation experienced a world that was undergoing huge political, social and economic changes (the end of the war and in some countries the return to a democratic system). We need only recall the processes of industrialisation and urbanisation or the migratory flows between and within European countries that took place in the fifties and sixties. But what it is particularly important to underline here is that that these processes went hand-in-hand with the extension of socio-economic rights. So, if we wanted to make use of the twofold term certainties/precarity, we could say that these processes went hand-in-hand with an expansion in the sphere of 'certainties', especially with regard to protection against the risks of sickness and disability that had affected previous generations. As is well known, from the sixties onwards, systems of universal welfare were instituted at varying speeds in European countries.

Finally, let us consider the generation of those aged 60-65 today. These subjects not only enjoyed a basic education but in many cases also had the opportunity of studying at school and university: in Italy only 5.2% of people in this age bracket do not have any school certificate whatsoever, as against 40.3% of the older generations; and 20% have at least a middle school certificate, as against 5% of the older generations. As far as their occupational status is concerned, most of these people worked in industry or the tertiary sector, as blue- or white-collar workers, but generally with some professional qualifications and with a high degree of job stability. Generally speaking, the people who were in their twenties in the period between the sixties and the seventies belong to a generation that in its youth and its mature adulthood lived through a historical phase in which the problematic features of the 'change of epoch' taking place in the post-war years also came to the surface. Alongside the consolidation of the certainties in social and work environments there was an increase in 'uncertainties' in the private sphere, both regarding the fate of one's marital relationship and the social and family condition of one's children. As a matter of fact, today's young people are increasingly exposed to growing 'precarity' both at work and in their personal relations.

In this third generation we find people who are currently facing a phase of transition to the third age, which can go on for many years. In this sense the label "elderly" is undoubtedly too narrow to characterise their condition. Many of these people continue to enjoy a state of health and engage in physical activities very similar to those of a mature adult: we might refer to them as "late-adults". Others, though continuing to enjoy good health, begin to experience a change in their social and family situation that leads to the assumption of roles that are much closer to those traditionally associated with old age: in this case we might adopt the definition "young elderly".

❖ *In your opinion, how does intergenerational solidarity between grandparents and grandchildren (and between grandparents and children in general) manifest itself and take form within the contemporary family?*

To understand the key features that characterise solidarity between grandparents and grandchildren today it is particularly important to keep in mind that as a consequence of the increase in life expectancy there has been a continuous growth in the number of situations in which the family scene is made up of three if not four generations. In this regard it is sufficient to note that the latest comparative survey conducted by SHARE (Survey of Health, Ageing and Retirement in Europe) has revealed that about one-third of people over 80 form part of a family that extends over four generations. This figure rises to between 40 and 50% in the majority of countries in northern and central Europe but falls to 20 to 30% in Austria, Switzerland and the Mediterranean countries, i.e. in those countries where there has been more rapid ageing but that at the same time are characterised by a limited birth rate and a tendency to have children at a later age. Alternatively, one could cite data from ISTAT (Istituto Italiano di Statistica), which shows how in Italy 98.2% of people under the age of 15 have at least one grandparent still alive (indeed, on average they have 3.1), and how 87.2% of those between the ages of 15 and 24 have at least one living grandparent. What this means is that people are much more likely to establish a relationship with their grandparents than in the past, and that this relationship extends over a considerable time, constituting one of the fundamental components of family relations.

At the same time, the last few decades have seen a marked increase in the employment of adult women or, in other words, married women with children. In Italy over the last 40 years this figure has risen from about 30% to almost 50%. This has led to a reshaping of the requirements and capacities for care-giving within families. It has meant, especially in countries like Italy with limited services for young children, that the elderly in their capacity as grandparents have been increasingly involved in the care of grandchildren. The latest *Multiscopo* survey by ISTAT, conducted in 2007, shows that, of the elderly that have at least one non-cohabiting grandchild, 85.6% take care of their grandchildren at least sometimes and that only 14.4% never take care of them. 24.4% of these grandparents take care of their grandchildren often, 15.7% in emergencies, 9.3% when their grandchildren are sick and 8.9% during the school holidays. What this means is that even when the care of grandchildren is not continuous, the presence of grandparents and the possibility of being able to rely on their support is nonetheless a fundamental factor, in that better use can be made of the services that are available, dealing with the gaps that they leave open due to the way they work (limited opening hours, closures during holidays, or unavailability in the case of children's illness).

The support that grandparents provide for their grandchildren and the closeness that derives from it is hugely important for both parties, above all in terms of affection, in that it enriches their relationship. This is particularly important if one considers the overall impoverishment in the quality of the social networks to which they each belong: the children, in that they increasingly tend to live in families in which they have only one sibling, or none at all, or in which there is only one parent; and the elderly, in that, when they retire, they often see a reduction not only in their social role but also in their network of friends.

But this support and this relationship are important for another reason: they introduce into the everyday life of the subjects in question two mutually reflecting temporal aspects – 1) a 'projectual' aspect for the grandparents, and for the grandchildren, 2) a different positioning in the history of the family (and perhaps also in history as a whole).

❖ *How does this role of grandparents impact on the relationship between elderly parents and adult children? Are there any gender differences in this regard?*

The supporting role that grandparents play in relation to grandchildren tends to reinforce the relationship between the grandparents and their own children as well. For example, the daily care provided to grandchildren has the inevitable effect that everyone sees each other and discusses how the day has gone. Besides that, the parents' support makes it possible for young women to work and thereby pursue their own biographical project. The help of grandparents increasingly takes the form of an opportunity offered to young mothers to continue to pursue their plans for a professional career.

In Italy, for example, it is no accident that research conducted over the last ten years has revealed a marked tendency for the different generations to live very close to one another. 5.5% of the people who have formed a family of their own live in the same apartment block as their mothers (11.6% within 1 kilometre; 11.2% in the same town/city). The figures in respect of people's fathers are only slightly lower: 4.8%, 10.5% and 11% respectively. Certainly, the regularity of contact and the support offered and received is facilitated by this physical proximity, but it is reasonable to assume that the proximity is actually desired and sought after precisely because these relations between the generations are regarded by both parties as a fundamental building block for the construction of their identity and for their social relations.

We should not forget, however, that there are significant gender differences at least in terms of the identity of the major care-giver. All the research shows that both the relations between parents and children and the relations between grandparents and grandchildren are more systematic in the case of women. Although the role of grandfathers is becoming increasingly significant, the process of caring continues to revolve around women. One reason for this is the fact that within the family, intergenerational support has been reinforced particularly in terms of care-giving (while support in terms of financial assistance has tended to become relatively less important), with a consequent expansion in women's contribution to family solidarity. This process of 'feminisation' is one of the most important characteristics of the change currently taking place in the forms of intra-family solidarity. Another type of change in intergenerational relations is emerging in certain contexts in Europe (above all where there is a strong tradition of family-based welfare): a reversal in the direction of family solidarity, or, in other words, no longer from young people to elderly people, but vice- versa. What is involved, however, is a partial reversal: for late-adults with parents of a very advanced age the traditional logic continues to prevail, which sees them as the principle care-givers of more elderly family members, but here too it is women who are involved to a greater extent.

❖ *Professor Rampazi, in your view do the elderly experience a degree of social uncertainty? And if so, why?*

The research that Carla Facchini and I have done over the years clearly demonstrates that this social uncertainty does exist, even though it takes on a range of different forms in the heterogeneous set of subjects normally labelled as "the elderly". The interesting thing is that while in the past the predominant aspect for the elderly seemed to be that of insecurity, for many of them today there are new situations of uncertainty, which have a number of features in common with those experienced by young people.

To clarify these claims I must first of all explain how "uncertainty" differs from "insecurity" and call to mind what the characteristics of uncertainty are in the experience of young people today. The principle feature of insecurity is fear, while the salient characteristic of uncertainty is doubt. Fear is paralysing, while doubt has an ambivalent effect: it can variously represent a brake on or a stimulus to action.

What does a person who feels insecure fear? Our hypothesis is that he/she is afraid of losing something that he/she possesses, or of not being able to compensate for the lack of "something" that he/she aspires to possess – something to which important characteristics of his/her personal and social identity are tied. When the possible loss - or lack - depends on factors that the subject definitely knows that he/she cannot control, an experience of precarity emerges (whether real or perceived) that can create a sort of paralysis of the will. Insecurity depends in part on the characteristics and the personal histories of the subjects in question, and in part on the type of guarantees that different social structures offer against the risks of physical, psychic, relational and economic difficulty.

What does a person in a situation of uncertainty have doubts about? Above all he/she has doubts about his/her capacity to make sense of his/her experience, to make the right life choices, to realistically evaluate the set of opportunities, risks and constraints that are present in his or her particular social context. The less defined and constricting the 'structuration' of the context in which one lives, the more generalised is the uncertainty: doubt implies the freedom to make choices, the meaning and consequences of which are not automatic, given a priori. In a situation of uncertainty one can be overwhelmed by the fear of not having sufficient resources to manage responsibly the liberty one has or that one thinks one has. But one can also be encouraged to exercise capacities of self-reflection, to define autonomously the direction of one's life course. In this sense, uncertainty contains within it both potential elements of insecurity and the possibility of devising strategies that enable one to control them.

The cultural framework of modern industrial society did not leave much room for the experience of uncertainty. The situations that were "not automatic" were viewed as exceptions, not the norm. Usually these coincided with particular key turning points or with the arrival at the thresholds of certain age brackets tied to the transition from one phase of life to the next. In particular, uncertainty was thematised as the principle characteristic of the moratorium conceded to young people as they sought to define their plans for adult life in line with their abilities and capacities. This type of project implies a choice between life courses whose meaning is clear-cut, whose evolution is foreseeable and whose unfolding is largely irreversible. In the past, the certainties entailed in such paths were guaranteed by a shared culture and by a structured and intelligible institutional system. Once a young person's doubts about what course was feasible and appropriate were resolved, he/she had no more to do than embark upon that path, knowing for certain that its overall direction and individual stages were implicit in the initial choice.

For this reason perhaps, uncertainty has traditionally been excluded from the analysis of the condition of adults and the elderly. By contrast, a great deal of attention has been given to the insecurity of certain categories of people, those subject to the risk of increased precarity in their lives. As far as this risk is concerned, the elderly have been and still are viewed as particularly exposed, because the process of ageing brings with it a potential loss of resources, in particular, resources connected with three important aspects of identity: the body, economic and social circumstances, and interpersonal relations.

The risk of increased precarity is linked with that of exclusion, in a logic of *disengagement*, and on this basis the idea emerges that old age coincides with a phase of life in which the time for plans is over. Generally, the event that symbolically marks the conclusion of this course is retirement. In the phase that opens up after withdrawal from active working life there is little to be done other than manage the loss of psychological, physical, and relational capacities associated with ageing. At the most, one can rely on the affection of the family to ward off loneliness and to preserve some form of projection into the future, however indirect: sharing the hopes of one's children and investing part of the time that remains in following the growth of one's grandchildren. There is no uncertainty in the idea of *disengagement*: becoming old is a natural, ineluctable fact. At most there may persist some element of unpredictability in respect of the greater or lesser rapidity with which a person's decline takes place after a certain age.

Today changes are occurring which are bringing new elements of uncertainty into the experience of youth and at the same time are undermining the system of certainties on which the elderly and late-adults have constructed their life courses. The de-institutionalisation taking place in late-modern societies goes hand-in-hand with a progressive individualisation/diversification in biographies. 'Non-automatic' life prospects are beginning to become a normal and generalised component of individual experience. On the one hand, there is an increase in the propensity to insecurity, tied to the economic and institutional crisis that has affected many western countries. On the other hand, there is an increasing emphasis, in the cultural imagination, on the idea that when the future is not automatic, not only will there be risks but also opportunities – opportunities, that is, for a reflective construction of oneself, a process which has the potential to develop over the course of one's entire life.

For young people this means having to formulate plans that are open to changes of mind and possible changes of direction. In the future, new opportunities and new constraints could emerge, difficult to foresee in the present, which will have to be managed with flexibility and inventiveness. Professional and affective equilibriums are becoming provisional: they have to be continually renegotiated and subjected to the test of doubt. The search for meaning in one's own life, typical of the moratorium phase, is beginning to manifest itself as a permanent challenge, which makes it ever more difficult to understand when the crisis of identity of youth comes to an end and when the transition to adulthood is brought to completion.

For people who are approaching retirement age, or who have already reached it, the changes taking place are creating the conditions for questioning the inevitability of the *disengagement* normally associated with that circumstance. Amongst these conditions there is, first, the fact that people age better and later than in the past: psycho-physical decline begins to become a handicap for active life well beyond the age of 60-65. Secondly, Western culture is beginning to accept the idea that the elderly, as well as being a problem, can also be a resource for society, because they have time, energy and capacities to devote to activities of various kinds: professional activity, voluntary work and caring for the family. In the third place, the growing instability in sexual relationships and, in general, the transformations in equilibriums within the family that to a varying degree characterise all western societies are changing role definitions for both the couple and the parent-child relationship.

As far as the couple is concerned, there is on the one hand an increasing risk of losing one's partner, given that in addition to the possibility of his/her death there is now the possibility of separation or divorce as well. On the other hand, there is also the possibility of legitimately seeking a new beginning either with one's original partner, when the children leave home, or with another partner, if the previous tie breaks down. The novelty is that all this now takes place at an age which in the past seemed to preclude any propensity towards a revitalisation of the relationship of the couple and, in particular, a full development of one's sexuality.

As far as the parent-child relationship is concerned, after the age of 60 there opens up a phase, of variable length, in which there is an overlapping of various roles. The subjects in question remain for a long time children of parents who live to a very old age. At the same time they continue to be actively present as parents in the life of their own children, who are struggling to achieve autonomy as adults. The consequence is that it is necessary for them to oscillate between expectations about their role that are not only very different but that evolve in ways and at-temporal rhythms that are difficult to foresee. This increases the uncertainty about one's place within the family and can generate ambivalent effects in the experience of late-adults/young elderly. On the one hand, there is an increase in the difficulty of managing everyday life, because it is necessary to constantly negotiate the boundaries of one's own role with other family members. On the other hand, one postpones the time when one's place within the family corresponds to that which is "typical" of an elderly person: increasingly less active and therefore progressively marginalised.

Obviously these changes do not affect all those over 60 to the same extent, and they do not produce the same effects on the experience of all the subjects in question.

❖ *Could you indicate more precisely what the effects of this new uncertainty are?*

As I have said, we are dealing with varied effects because they depend on the characteristics of the social system of which the subjects are a part, on their personal resources and on the type of biographical course that they have behind them.

As far as the characteristics of the social system are concerned, uncertainty tends to manifest itself above all in terms of personal insecurity in contexts where there is more weakness and instability in the labour market and where the welfare system offers limited protection against the risks of unemployment and poverty. It is necessary, however, to keep in mind the risk of affective precarity and loneliness, which is particularly high in countries in which traditional family ties have unravelled to a greater degree. These characteristics impact above all on the type of certainties offered to young people, but they also impact on the less young, who may be directly affected by the dismantling of institutional protection and by the precarisation of family relations. The certainties of the elderly, however, can also be undermined indirectly, when their condition is in some way influenced by the difficulties that the younger generations have in achieving independence and stability. We have coined the term "reflected uncertainty" to describe this phenomenon.

The level of reflected uncertainty is particularly high in countries where there continue to be strong ties of intergenerational solidarity, supported by a family-based type of welfare system. As Carla Facchini has noted, in many countries the direction of solidarity between generations is undergoing a reversal of direction *vis-à-vis* the past, at least in respect of the type of help that young people expect from adults and from the elderly. This means, for example, that in the absence of effective policies for ensuring economic independence and housing for the young, recourse is made to the resources of the respective parents and grandparents. In this way parents and grandparents risk losing substantial economic resources and having to subordinate the use of their own time and energies to the needs of children and grandchildren. The instability of young people's family lives may also impact on the life of the older generations. We need only think of those young people that leave home at an increasingly advanced age and then sometimes return home to seek further support when they remain single, not infrequently in financial difficulties and/or with young children to look after. This can lead to parents having to wait a long time before divesting themselves of responsibility for the psychological support and care of their children. One consequence of this, for example, is that the parents of these young people are not able to predict whether or when they will become grandparents, whether or when they will be able to reconsider their relationship with their own partner or whether or when they will be able to stop worrying about the happiness of their children, so as to be able to concentrate primarily on themselves.

For the current group of late-adults/young elderly, however, reflected uncertainty may also depend on the type of solidarity that ties them to their own parents, still living and often no longer self-sufficient on account of their advanced age. In contexts where welfare systems do not guarantee adequate home-based and rest-home services for these "old elderly", it is their children, especially their daughters, often over the age of sixty themselves, who have to take on the caring tasks. These are extremely onerous both in financial terms and in terms of time, effort and emotional stress. The weight of this commitment can translate into a deterioration in the material and relational resources of these care-givers, to the point that relationships with their partners and children are jeopardised.

Systemic factors, however, do not completely account for the different ways in which uncertainty manifests itself in the biographies of the different categories of the elderly. As we have said, the differences also depend on the personal histories of the subjects in question. These histories are the product partly of the capacities of each person, partly of the resources and constraints connected to his/her original social class, and partly of the manner in which he/she has been exposed to the transformations that Western societies have undergone in the course of the twentieth century. As Carla Facchini has already underlined, the generic label of "the elderly" conceals a number of age cohorts and a number of generations. Membership of particular age cohorts impacts on people's psycho-physical resources and on their position in the system of family roles. Membership of a generation, interacting with economic and cultural status, influences the type of certainties a person enjoys and the way in which he/she interprets the new propensity towards uncertainty that is manifesting itself today.

In other words, having guarantees in terms of health and economic well-being is a necessary but insufficient condition for being able to identify opportunities for biographical construction, once one has reached the age of full adulthood. It is also necessary to have adequate cultural resources and a propensity towards the reflective construction of oneself, developed in the preceding phases of one's life. In this way, for example, considering various generations of Italians over the age of sixty, we have noted that such a propensity is greater in those who, as well as enjoying a medium to medium-to-high socio-cultural status, belong to the post-war generation. These subjects underwent a kind of pre-socialisation to the culture of uncertainty.

As young people, in the years between the sixties and seventies, they experienced the questioning of the system of certainties typical of modern industrial societies. As adults, they witnessed the progressive destructuration of this system, which has led to the individualisation of life courses. In the face of old age, they know that they will be able to remain active for a long time, in good health and financially reasonably secure. This allows them to think that they still have opportunities for personal development. The future remains open, even though they do not always know for certain if and how these opportunities will take concrete form, whether they themselves will be capable of exploiting them and, if so, for how long they will be able to do so.

❖ *You speak of a possible openness towards the future. Could you explain exactly what form the relationship between the elderly and time takes today?*

For the reasons already outlined, I do not think it possible to speak in general terms about a particular relationship between the elderly and time. What one can do is consider the different temporal horizon that is opening up for subjects no longer young, positioned in various contexts and with different resources and variegated histories behind them. Within this multiplicity of situations we can identify a certain number of typical-ideal models which, nonetheless, do not exhaust the wide array of cases that are present on the scene today.

The first model re-proposes the temporality implicit in the traditional logic of a project that gives rhythm to the various phases of life. This is the logic that, as I have already indicated, lies behind the idea of disengagement, once adult life has come to an end: in the present there is nothing more to construct, because the project, for better or worse, has been brought to completion. One's identity resides in the past, which at times projects itself into the present to the point of engulfing it. The future does not exist, other than in the form of the certainty of decline, which one prefers not to think about. The dimension of uncertainty, when it is present, for the most part assumes the semblance of insecurity, above all in the case of the "old elderly" and, in general, of those who are exposed to a greater extent to the risk of poverty, disability and loneliness.

This type of orientation can also be found in some late-adults or young elderly, who currently enjoy a series of certainties about their economic, physical and affective well-being. These people have internalised the traditional logic of disengagement and are experiencing a sort of temporal interval, freed from adult responsibilities and from the constrictions of old age.

In this interval the present is not just empty time, to be filled up in some way, but is rather a reserve of *free time* to be taken advantage of. It is a time for making leisure plans and/or for caring, in the framework of relationships with partners, family members and friends. The idea is not so much one of personal construction as one of reaping the fruits of what one has become, thanks to the life course one has completed.

One could say that for these people the dominant aspect is that of certainty, if it were not for the fact that some of them are to varying degrees exposed to reflected uncertainty. The level of uncertainty in this case depends on the type of solidarity that the subjects in question believe is necessary to show towards their own children, now adults, faced with the growing risks of insecurity in employment and affective instability. It needs to be kept in mind, moreover, that for women especially, reflected uncertainty potentially has a twofold origin: the difficulties of their children, on the one hand, and, on the other, the precarity of their parents, by now quite elderly. The consequence of this is that they are substantially limited in their ability to manage their own financial, temporal and relational resources for themselves. These limitations result in the transformation of the potential free time that they could enjoy into an excess of bounded time, in the service of family solidarity.

Alongside these various manifestations of the "traditional" temporal model we also find a second, more innovative model. This manifests itself in some categories of late-adults or the almost-elderly who, as I have indicated, have been pre-socialised to the new forms of uncertainty and who, as well as enjoying good financial and health prospects, also have substantial cultural resources.

In these circumstances such subjects are able to live out the long transition towards old age as a period dedicated to self-discovery. The past is behind them, and the future is a further more or less extensive segment of their biographical life course, open to construction. What is involved is a brief future, or an extended present, similar to that which characterises the temporal perspective of many young people. The physiognomy of this future is defined as one goes along: it depends on the capacity of subjects to take advantage of novelties, to exploit the unexpected events that lie hidden within everyday life. It might be possible, for example, to take up projects that have been put aside in the past because of work or family problems, at times effecting an out-and-out restructuring of one's biography. Or one might undertake new activities, of a professional nature or in the field of voluntary work. One might even discover hitherto unknown artistic abilities or plan a whole new beginning to one's emotional life.

But within this innovative orientation too, the phenomenon of reflected uncertainty can produce interference in the two forms indicated above.

Here too, more or less severe restrictions arise on the liberty of late-adults/the almost-elderly. It is very common for a contradiction to arise between the aspiration to explore new courses and the need to fulfil responsibilities taken on in the past. For some, this bind is experienced as a lack of control over their life, and it can translate into an experience of insecurity. For others, reflected uncertainty is a further manifestation of non-automaticity, to be managed by seeking a temporary balance between aspirations and reality, in the expectation that sooner or later the situation may change.

References

The authors have conducted research on the transition to the condition of being elderly. See in particular:

- Facchini, C. & Rampazi, M. (2006). *Generazioni anziane tra vecchie e nuove incertezze [The Elderly Generations between Old and New Uncertainties].* Rassegna Italiana di Sociologia 1: 61-90.
- Facchini, C. & Rampazi, M. (2008). *Generazioni ad un passaggio d'epoca. Certezze e precarietà nei racconti degli ultrasessantenni [Generations in a change of epoch. Certainties and Precarity in the Accounts of the Over 60s].* in Ruggeri, F. (ed.). *La memoria del futuro. Soggetti fragili e possibilità di azione* [The Memory of the Future. Fragile Subjects and the Possibility of Action]. Milan: Angeli, 113-137.
- Facchini, C. and Rampazi, M. (2009). *No Longer Young, not yet Old. Biographical Uncertainty in Late-adult Temporality. Time & Society* 18.2/3: 351–372.

Other references

- Brückner, H. & Mayer, K. U. (2005). *De-Standardization of the Life Course: What It Might Mean? And If It Means Anything, Whether it Actually Took Place,* in Macmillan, R. (ed.). *The Structure of the Life Course: Standardized? Individualized? Differentiated?* Advances in Life Course Research 9: 27-54. Amsterdam: Elsevier.
- Caradec, V. (2004) *Vieillir après la retraite: approche sociologique du vieillissement.* Paris: PUF.
- Cavalli, A., Cicchelli, V. & Galland, O. (eds.). *Deux pays, deux jeunesses? La condition juvénile en France et en Italie.* Rennes: PUR.
- Côté, J. (2000). *Arrested Adulthood. The Changing Nature of Maturity and Identity.* New York: New York University Press.
- ISTAT (2001). *Parentela e reti di solidarietà, Indagine multiscopo sulle famiglie. Aspetti della vita quotidiana.* Anno 1998, Rome.

- Leccardi, C. (2005). *Facing Uncertainty. Temporality and Biographies in the New Century, Young. Nordic Journal of Youth Research* 13.2: 123-146.
- Maltby, T., De Vroom, B., Mirabile, M.L. & Overbye, E. (eds.) (2003). *Ageing and Transition to Retirement. A Comparative Analysis of European Welfare States.* Aldershot: Ashgate.
- Mannheim, K. (1952). *Essays on the Sociology of Knowledge [Edited by Kecskemeti, P.]* New York: Oxford University Press.
- Mayer, K. U. (2004). *Whose Lives? How History, Societies and Institutions Define and Shape Life Courses, Research in Human Development* 1.3: 161-187.
- Naldini, M. (2002). T*he Family in the Mediterranean Welfare States.* London: Frank Cass.
- Nowotny, H. (1988). *From Future to the Extended Present. Time in Social Systems,* in Kirsch, G., Nijkamp, P. and Zimmermann K. (eds.). *The Formulation of Time Preferences in a Multidisciplinary Perspective,* Aldershot: Gower, 17-31.
- Saraceno, C. (ed.) (2008). *Families, Ageing and Social Policy. Intergenerational Solidarity in European Welfare States.* Cheltenham, UK: Edward Elgar.
- Vincent, J. A., Philipson, C. R. & Down, M. (eds.) (2006). *The Futures of Old Age.* London: Sage (in association with the British Society of Gerontology).

2.3 Ambivalences, Conflicts and Solidarities Within the Family Today

Interview with Ariela Lowenstein
Department of Gerontology / Center for Research & Study of Aging,
Faculty of Welfare and Health Sciences at University of Haifa

[Editorial note: intergenerational solidarity plays a crucial role today in social relations and, in particular, in relations within the family: indeed, the family has become the privileged locus of expression for this solidarity. Within the contemporary family there are, in fact, a multiplicity of types and forms of support that manifest themselves between the various generations, young and less young. Nonetheless, it should be kept in mind that these forms of intergenerational solidarity (or their possible absence) can also, in some cases, generate conflicts, a sense of guilt and ambivalence – both in those who offer them and in those who receive them. As far as the particular phenomenon of ambivalence is concerned, which forms the central theme of the following interview with Ariela Lowenstein, it is important to provide a definition beforehand. In the present context the term ambivalence is intended to refer above all to those situations and specific circumstances characterised by oscillation between opposing attitudes and approaches. Because of uncertainty, which goes hand-in-hand with ambivalence, choices and decisions become particularly difficult. We have asked Professor Ariela Lowenstein to offer her views on these questions.]

❖ **Can you discuss the concept of intergenerational ambivalence, both from a theoretical and from an empirical perspective?**

The intergenerational ambivalence perspective to the family as a system stems from the modern, or rather postmodern era of the twenty-first century. This era is characterised by pluralism and multivalency, thus putting the individual in constant existential dilemmas of choosing between competing meanings. This chaos of meaning causes a psychological experience of ambiguity and ambivalence, characterised by conflicting feelings: the need for liberation on the one hand, and the fear of alienation on the other. Conflicts and contradictions are not only typical of the individual at the micro level, but also characterise society as a whole at the macro level. This assumption is the basis for the concept 'sociological ambivalence', first formulated by Merton/Barber (1963). They define sociological ambivalence as incompatible normative expectations of attitudes, beliefs and behaviour.

Family researchers have integrated such perspectives dealing with ambivalence at the personal and interpersonal level with the theories dealing with ambivalence at the larger social scale (sociological ambivalence) to formulate the concept of intergenerational ambivalence. Generally, intergenerational ambivalence can be defined as simultaneously held opposing feelings or emotions that are due in part to countervailing expectations about how individuals should act. More specifically, intergenerational ambivalence is viewed as a concept constructed at two different structural levels: the macro and the micro level. As such, its definition should capture these two levels. Thus, according to Luescher 'intergenerational ambivalence' reflects contradictions in parents' and adults' offspring relationships in two dimensions: "at the level of social structure in roles and norms - the macro level" and "at the subjective level, in terms of cognitions, emotions and motivations - the micro level".

Following these conceptual definitions of intergenerational ambivalence, initial attempts were made to define the concept operationally. Luescher's model captures the two dimensions of ambivalence: the structural (macro) dimension and the inter-subjective (micro) dimension. Each dimension is represented by two poles: the structural dimension is represented by the poles of reproduction versus innovation and the inter-subjective dimension is represented by the poles of convergence versus divergence.

At the macro level, each family system can be seen as a sociological institution, characterised by a specific structure, as well as by norms and procedures, which represent the values and conditions of the larger society in a specific cultural era and geographic place of living. These institutional values and conditions are, on the one hand, reproduced through the acting out of relations (solidarity, captivation) by family members. On the other hand, these values and conditions could also be modified (emancipation, atomisation), thus leading to innovations. Hence, reproduction and innovation are the two poles in which the family is realised as a social institution. In this model, these two poles represent structural ambivalence. If one scores highly on both poles, then one is viewed as ambivalent in the structural sense, since the two poles represent opposite themes.

At the micro level, each family can be conceived as an emotional, intimate unit, which contains the potential for closeness and subjective identification, thus reinforcing similarity between the children and their parents. Similarity and closeness are psychologically gratifying, on the one hand, but on the other hand they can also be experienced by the family's member as a threat to individuality. Thus, the family members are motivated to keep the unit's cohesion (convergence), but on the other hand they strive for separation and individuality (divergence). Hence, Luescher sees convergence and divergence as two poles representing inter-subjective ambivalence. If one scores highly on both convergence and divergence, then one is viewed as ambivalent at the micro level.

An altogether different way that ambivalence can manifest is using feelings of guilt as a key concept representing ambivalence.

❖ *In this sense, how can we define "guilt" and what is its role in the theoretical conceptualisation of ambivalence?*

We can view guilt as belonging to what Lazarus/Lazarus outlined as 'the existential emotions': "Anxiety-fright, guilt and shame are existential emotions because the threats on which they are based have to do with meanings and ideas about who we are, our place in the world, life and death and the quality of our existence. We have constructed these meanings for ourselves out of our life experience and the values of the culture in which we live and we are committed to preserving them". They see guilt as an emotion experienced when one feels personal failure, as a result of a moral lapse. They believe that guilt can be regarded as a kind of anxiety.

Their existential view of guilt is especially relevant when relating to guilt that care-givers feel towards their elderly parents. Since the elderly are close to the end of their life, being close to them is certainly bound to induce existential (death) anxiety. The care-giver, thinking about the institutionalisation of his/her parent (or any other close relative) cannot help but think about death, consciously or unconsciously. Other authors, such as Wentzel, even assume that one of the reasons care-givers find the decision to institutionalise their elders so difficult is because it makes the care-givers think of their own death.

Lazarus/Lazarus' view of guilt as connected to morality provides a theoretical explanation as to why care-givers often feel guilty towards their elderly parents, when it seems to them that they are not providing the best care possible. This violates the moral code that one should not neglect his parents when they grow old. Some articles describe this link between guilt feelings and a sense of moral misdeed and show that care-givers indeed feel guilty when they believe they haven't done the right thing morally, using personal stories.

A slightly different conceptualisation of guilt views it not as an existentialist, but mostly as social and interpersonal. This way of looking at guilt is concerned with a deed that has violated certain social norms. Another central aspect of guilt is the interpersonal aspect. In guilt, as in other emotions that are typically related to those close to us, our relationship to our intimates is of central importance. People's descriptions of guilt-inducing situations often highlight neglect of a partner or a failure to live up to expectations. This view explains why guilt is often experienced in intergenerational family relationships, since these relationships are usually very close and intimate, and characterised by high expectations of support in situations of sickness and disability.

Empirical data show, for example, that feelings of guilt are common during the placement of one's parents in a nursing home; it seems that institutionalisation of the elderly basically induces one major emotion on the part of the care-giver: guilt. However, the picture is more complex. Some empirical studies show that the institutionalisation process is actually accompanied by ambivalent feelings on the part of the care-giver: on the one hand, feelings of guilt and grief, but on the other hand feelings of relief, that the burden of the care has been lifted (Riddick/Cohen-Mansfield/Fleshner/Kraft, 1992; Ryan/Scullion, 2000). In sum, guilt feelings most often go with ambivalence, with moral considerations and contradicting practical considerations. This is one of the reasons why guilt is an emotion representing ambivalence.

❖ *Why can guilt be considered a key concept for representing ambivalence?*

On a general level, thinking about the different situations when one feels guilt within the family, one of the main characteristics of all of these situations is a sense of ambivalence, a sense that one is torn between two or more options, without being able to feel he has chosen the correct one. Thus, when one chooses a specific option, but does not feel he has done the right thing, guilt often arises. When speaking about family relations, guilt is bound to arise in some specific situations, which can be shown using the heuristic model:

- when a family member uses atomisation, and separates conflictually from other family members, he is likely to feel guilty, since in every family there is a side that wishes for solidarity and closeness, and wishes to please other family members;
- there are times when a family member uses captivation and does what most family members want, although he may think the right or moral decision should have been different. This is another situation which may well give rise to guilt feelings.

Generally, modes of divergence are likely to increase feelings of guilt, as opposed to modes of convergence. Thus, feelings of guilt may represent one aspect of the inter-subjective dimension of ambivalence.

❖ *Can you give us an example of "ambivalence" on an empirical level?*

A study conducted with Rachman, more than a decade ago, was designed to examine the reasons for the decision to institutionalise an older parent, comparing the city and the kibbutz in Israel, and to analyse its impact on intergenerational family relationships. The hypothesis was that the following four factors would be the main causes for institutionalisation: 1. the burden of care; 2. the exchange relationships between adult children and older parents; 3. the role of children; 4. the role of the formal support systems. The research aimed to find out how these four factors influenced the decision to institutionalise and the relationship between the family members.

It was assumed that this process would generate more conflict in the city, since the care-giving burden (economical, physical and emotional) is higher and multi-faceted there, while in the kibbutz the burden is much lighter and mostly emotional. Another difference between the city and the kibbutz, which makes the institutionalisation in the kibbutz a somewhat smoother process, is the formal service system, which was at the time much more readily available in the kibbutz than in the city. Since formal support systems were found to contribute greatly to an effective placement of an elderly in an institution, it is reasonable to assume that the support system in the kibbutz would significantly ease the process of institutionalisation compared to the city, where it is much less accessible and provides less formal help and support. This was confirmed. The main idea of this study was to show how the kibbutz's norms support institutionalisation, especially since in the kibbutzim studied, the nursing homes were part of the kibbutz, making it legitimate and thus diminishing guilt and the feeling of ambivalence, while the picture in the city is reversed.

The theories and findings concerning the role of children and the role of the formal support systems are most relevant to an analysis of the norms and expectations concerning institutionalisation in the city versus the kibbutz. The social norms governing children's behaviour towards their elderly parents in the city are based on the concept of "filial responsibility". This concept means that children feel a personal obligation to ensure their elderly parents' well being, trying to protect them and care for them. These views and attitudes are expressed in certain behaviours towards the elderly, such as: shared household arrangements, helping with tasks, keeping in touch and providing emotional support. These norms have an impact on the decision to institutionalise an elderly parent. Although in the city the instrumental and emotional burden is high, over 40 per cent of the care-givers doubted and speculated more than half a year before starting to check the possibilities of placement in an institution. This confirms the children's high feelings of obligation and responsibility towards their parents. Seventy per cent of the children in the city said that the institutionalisation took place when they had no other choice, since there were not enough support services.

In sum, children in the city find themselves in a complicated situation concerning norms about institutionalisation: on the one hand, they feel responsible towards their parents, and therefore they try to keep them at home for as long as possible: on the other hand, the instrumental and emotional burden as well as the lack of formal support systems makes it, in certain situations, almost impossible to do so. Thus they find themselves in an emotional and practical conflict, exhibiting feelings of ambivalence.

The Israeli kibbutz was still at the time of the study a unique kind of community, characterised by a full partnership of its members in all areas of life: economics, health, education, housing, etc. Each member of the kibbutz emotionally experiences the kibbutz's society as his extended family. Thus, in the ideological-social structure of the kibbutz, obligations to the community are equal to obligation to one's family. This makes the children less personally obligated to provide instrumental support to their parents. They tend to take less responsibility for their elders, since they know the kibbutz will do so. In many of the veteran kibbutzim, for example, nursing homes were built within their grounds to serve their elderly members "at home".

This is one difference between institutionalisation in the kibbutz versus the city: the elderly moving to a nursing home in the city have to adjust to a basic change of environment, moving from home to a 'total' institution. By contrast, in the kibbutz the elderly move from their home to a nursing home in the same environment, a move which is less traumatic. Another difference between institutionalisation in the city versus the kibbutz is the decision itself. In the city, the decision to institutionalise is taken by the close family, and often causes conflicts between siblings and between them and the elderly parent. This way, responsibility for the decision rests on the whole family. In the kibbutz, the situation is totally different. The family is not alone in its decision, but the kibbutz's formal support system takes much of the responsibility. When the functional situation of the elderly requires constant formal help, the kibbutz's support system decides that the elderly individual has to move to a nursing home. This is an economic decision, because in this way there is no need for a private nurse in the elders' house and the children can go back to their productive function in the kibbutz. Thus in the kibbutz the family is able to share the decision with others, thus diminishing feelings of responsibility and guilt. Badgwell found that sharing the decision with other family members helped to reduce feelings of guilt, as did involvement in local support groups. In fact the kibbutz is sort of a local informal support group, helping the members ease the emotional burden, which is part of the institutionalisation process.

❖ *According to the studies and research you carried out, how different are the patterns of intergenerational solidarity, conflict and ambivalence observed across several societies that differ in welfare provision and family traditions?*

The data related to the OASIS[1] project suggest that the majority of respondents in all the five countries considered - Norway, England, Germany, Spain and Israel - reported strong and positive emotional solidarity (affective-cognitive solidarity) between adult children and their old parents, whereas the negative emotional feelings (conflict and ambivalence) were low. These findings confirm, in cross-cultural contexts, that the extended family today has maintained cross-generational cohesion with some conflict as well as some ambivalent feelings (Pillemer/Luescher, 2004). The data thus support the more recent perspective of the solidarity-conflict model. Further study of the balance between solidarity and conflict is therefore needed, and a further exploration of ambivalence is also warranted, focusing on how it emerges in family relationships.

The similarities as well as the differences found between the countries in the various dimensions of solidarity-conflict and ambivalence may reflect variations in family norms and behaviour patterns, as well as traditions of social policy in the participating countries. This heterogeneity can be attributed to historical trends over the last century. In linking the testing of solidarity-conflict and ambivalence at the micro level of individuals and families to the macro perspective of the cross-national study, unique idiosyncratic historical and familial developments in the context of the countries involved must be taken into account. The higher rates of close parent-child relationships found in Israel may be closely related to the country's recent history and geopolitical situation. However, the higher rates of conflict might reflect a culture where open communication between generations is encouraged. Similarly, the apparent generation gap between current cohorts of older parents and their adult children in Germany may be related to the polarisation along generational lines of traditional/radical attitudes that occurred in the 1960s. In Spain, findings of relatively low rates of close parent-child relationships, contrary to expectations, may be due to rapid modernisation (reflected, for example, in low fertility rates). Younger generations are more exposed to this process, and are better educated and better-off than their parents. This could result in the emergence of a significant generation gap.

---

[1] The OASIS (Old Age and Autonomy: the Role of Service System and Intergenerational Family Solidarity) research project was funded within the 5th Framework Programme of the European Community. The overall goal was to discover how family cultures and service systems support autonomy and delay dependency in old age, so as to promote quality of life, and improve the bases for policy and planning. See *http://www.oasis-project.eu/*.

Participating OASIS-countries also represent different contexts and opportunity structures for family life and elder care. They are confronted by similar challenges in this area, but are inclined toward different solutions. Of particular interest is that Germany and Spain are welfare states that tend to favour family responsibility and play a subsidiary role (Germany) or even a residual (Spain) role. Both countries lay down legal obligations between generations but have relatively low levels of social care services, although they may have high levels of medical services. By comparison, England and Norway have individualist social policies, no legal obligations between generations, and higher levels of social care services. Younger generations there find it easier to combine work with family obligations than in Germany or Spain. The mixed Israeli model is illustrated by legal family obligations, as in Spain and Germany, with high service levels, as in Norway. The solidarity-conflict model was especially useful in evaluating the strength of family relationships in the different societies. However, conceptually, the model does not claim to capture the entire complex and diverse picture of late-life family relations, as acknowledged by Bengtson et al. This is especially true at points of transition along the life course, such as the failing health of older parents or the changing needs of working care-givers, when more negative and/or ambivalent feelings may surface.

The 'operationalisation' of ambivalence was in its infancy when the OASIS study started, and we used what was suggested by their originators – Luescher and Pillemer. Actually in OASIS, ambivalence was best captured through the qualitative data. Solidarity-conflict was measured mainly by quantitative data over the years but, as Giarrusso, Silverstein, Gans and Bengtson indicate, there is an on-going effort to refine the items measuring solidarity and conflict which in the years since the study started makes the measuring instrument a 'gold standard' for studying and assessing intergenerational family relations. Using both quantitative and qualitative methods of data collection and the triangulation of data bases is recommended in order to further address and examine these different concepts.

Recent research attempting to operationalise ambivalence and validate it by capturing its individual and structural dimensions in central life course transitions has been published (Pillemer/Luescher, 2004). The accumulation of additional empirical evidence would facilitate further theorising and identify the ways in which it emerges in family relationships. In this respect, some answers are given, but new, intriguing questions and issues arise: does ambivalence complement solidarity and conflict as a form of family relationship, especially during periods of transition? Is there a need to further explore the three concepts - solidarity, conflict and ambivalence - in additional cross-national and cultural idiosyncrasies, to better validate their accuracy in explaining parent-child relations in adulthood?

References

- Badgwell, N. (1986-1987). *'Hardest Decision',* Modern Maturity 29.6: 82-86.
- Bengtson, V. L., Giarrusso R., Mabry B., Silverman, M. (2002). *'Solidarity, Conflict and Ambivalence: complimentary or competing perspectives on intergenerational relationships?'* Journal of Marriage and the Family 64: 568-576.
- Giarrusso, R., Silverstein, M., Gans, D. & Bengtson, V. L. (2005). *Aging parents and adult children: New perspectives on intergenerational relationships,* in Johnson, M.L. Bengston, V.L. Coleman, P.G. & Kirkwood T.B.L. (eds.) *Cambridge Handbook on Aging.* Cambridge: Cambridge University Press, 413-421
- Lazarus, R.S., Lazarus, B.N. (1994). *Passion and Reason: making sense of our Emotions.* New York: Oxford University Press.
- Luescher K. (1999). *Ambivalence: a Key Concept for the Study of intergene,rational Relationship,* in S. Trnka, Family Issues between Gender and Generations, Seminar report, European Observatory on Family Matters, Wien.
- Luescher, K. (2004). *Conceptualising and uncovering intergenerational Ambivalence,* in K. Pillemar and K. Luescher (eds.) *Intergenerational Ambivalences: new Perspectives on Parent-Child Relations in later Life,* Oxford: Elsevier, 23-62.
- Lowenstein A., Rachman, Y. (1995). *Factors involved in the Process of the Decision to institutionalise the Elderly in a nursing Home in the City versus in the Kibbutz and the Way these Factors influence the Relationships in the Family,* Gerontologia, Israel Gerontological Society.
- Lowenstein, A. & Daatland, S.O. (2006). *Filial norms and family support to the old-old (75+) in a comparative cross-national context (the OASIS study),* Ageing and Society 26: 1-21.
- Lowenstein, A. (2007). *Solidarity-conflict and ambivalence: Testing two conceptual frameworks and their impact on quality of life for older family members,* Journal of Gerontology Social Sciences 62B: 100-107.
- Lowenstein, A., Katz, R. & Gur-Yaish, N. (2008). *Cross-national variations in elder care: Antecedents and outcomes,* in M.E. Szinovacz and A. Davey (eds.) *Caregiving contexts: Cultural, familial and societal implications,* New York: Springer Publishing, 93-114.
- Merton, R. K. & Barber, E. (1963). *Sociological Ambivalence,* in E. Tityakian (ed.) *Sociological Theory: Values and socio-cultural Change.* New York: Free Press.
- Pillemer, K., Luescher, K. (2004). *Intergenerational Ambivalences: new Perspectives on Parent-Child Relations in later Life.* London/New York: Elsevier.

- Rappaport, A., & Lowenstein, A. (2007). A *possible innovative association between the concept of intergenerational ambivalence and the emotions of guilt and shame in caregiving*, European Journal of Aging 4.1: 13-21.
- Riddick, C.C., Cohen-Mansfield, J., Fleshner, E., Kraft, G. (1992). *'Caregiver Adaptations to having a Relative with Dementia admitted to a nursing Home',* Journal of Gerontological Social Work 19.1: 51-76.
- Ryan, A.A., Scullion, H.F. (2000). *'Nursing Home Placement: an Exploration of the Experiences of Family Careers',* Journal of Advanced Nursing 32.5: 1187-1195.

2.4 Intergenerational Solidarity and EU Citizens' Opinions: Some Indications for Policy Making

Francesco Belletti
Forum delle Associazioni Familiari / Centro Internazionale Studi Famiglia

1. Generations in society and family

Intergenerational solidarity has always been one of the main responsibilities of family life, but it is also one of the fundamental dimensions for social cohesion as well. In this twofold perspective, solidarity between generations has been crucial in building welfare systems at national and local level, and today a new balance between the specific contribution by families and state intervention seems necessary, especially in the light of the so-called "demographic transition" of the last few decades.

> *"Through its Green Paper Confronting Demographic Change, published in March 2005, the Commission initiated a debate on the need to strengthen solidarity between the generations. [...] The debate which then started in Europe on the subject of demographic ageing has added to this perspective. It has become clear that the balance in European societies rests on a set of inter-generational solidarity relationships which are more complex than in the past. Young adults live under their parents' roof for longer, while, increasingly often, the parents have to support dependent elderly people* (First European Quality of Life Survey 2003, European Foundation for the Improvement of Living and Working Conditions). *The resulting burdens are borne mainly by the young or intermediate generations, and generally by women. Equality between men and women, and equal opportunities more generally, would therefore appear to be key conditions for the establishing of a new solidarity relationship between the generations"*
> (Introduction of the Communication from the Commission to the European Parliament, the Council, the European Economic and Social Committee and the Committee of the Regions, "Promoting Solidarity between the Generations", Brussels, 10.05.2007).

This formal and official declaration from the EU resulted from a gradual scientific awareness of the importance of the intergenerational aspects of public and private solidarity, such as the prescient warning from Pierpaolo

Donati, issued in the 1991 CISF Report on the Italian Family (for more details, see *http://www.cisf.it*):

> *"In order to handle the consequences of the demographic transition we need more than a mere 'pact between generations', considered as age groups confronting each other in the public arena and competing for a present or future share of resources (i.e. working opportunities or financial resources for pensions); we need rather to define which criteria are linking the different age groups and connecting the decisions about the present and the near future, not only in society, but also in families as well. This general framework - the linking criteria - preliminary to the specific intergenerational pact, can be defined as an alliance between family and society [...] since families - and generations - are relational goods"* (Donati, 1991: 404).

In other words, Donati was stressing the importance of the intergenerational dimension of family relations and in society as one of the main elements necessary to build social cohesion and solidarity in families and in society.

2. Ageing society and intergenerational solidarity

The relevance of intergenerational solidarity in European societies is strongly stressed also by NGOs lobbying for elderly people, avoiding, in such a way, a sort of intergenerational competition for scarce public resources.

> *"In our view, enhanced solidarity between generations can play a key role in developing fairer and more sustainable responses to the major economic and social challenges that the EU is facing today. Our society needs to become more inclusive to allow everyone to get involved whatever their age, gender, ethnic origin, skills and ability. Action is also urgently needed to ensure a fairer redistribution of resources, responsibility, and participation, and to develop greater cooperation between generations in all social and economic spheres. It is important in today's context to maintain a high level of solidarity in our social protection systems given its proven shock absorber effect during economic crises. Public authorities should develop holistic and sustainable policies supporting all generations, facilitate access to adequate income and to affordable and quality services, particularly housing, education and health for people of all ages, and foster exchange of good*

> *practice and mutual learning between different generations. Engaging migrant and minority communities in intergenerational solidarity initiatives together with majority communities is crucial for breaking down harmful stereotypes, bringing communities closer together, dispelling myths and creating public space for dialogue. Raising awareness of creative social solidarity initiatives developed by migrant and minority communities, including women's organisations, is particularly important"*
> *(AGE Platform Europe, 2010)* [1].

This approach can also support a positive social representation of elderly people ('Active Ageing'), fighting against negative stereotypes of dependency and economic and social burdens:

> *"Demographic ageing is strongly affecting the relationships among generations and the way European societies function. Rather than focusing on the negative challenges of ageing, such as its impact on the increased pension and health care expenditure or on the shrinking labour force, demographic reality should be looked at as an opportunity, which can bring solutions to many current economic and social challenges, but therefore requires a new assessment and reworking of several economic and social policies within society"*
> *(AGE Platform - Press pack: Intergenerational Solidarity, 2010).*

3. Elderly people as a resource: information from Eurobarometer

Social and family policies at local and national levels are trying to promote intergenerational solidarity among generations as social groups, in the public area, but in most nations the strongest flow of mutual help among generations is found within the family. So it is important to consider how people actually perceive the relationships between generations, and the role of elderly people in this reciprocal and bidirectional exchange of resources. A recent survey provides information on social orientations towards intergenerational solidarity and the role of elderly generations. The *Flash Eurobarometer* "Intergenerational Solidarity" (Flash No. 269) fieldwork was conducted between 20 and 24 March 2009. Over 27,000 randomly-selected citizens aged 15 or over were interviewed in the 27 EU Member States. Interviews were predominantly

---

[1] "Intergenerational Solidarity: the Way forward. NGOs coalition calls for 2012 to become European Year of Active Ageing and Intergenerational Solidarity", from the Joint Press Release in preparation for the Second European Day of Solidarity between Generations, 29th April 2010, Logrono, Spain. See *http://www.age-platform.org*.

carried out via fixed-line telephone, reaching some 1,000 EU citizens in each country. Parts of interviews in Austria, Finland, Italy, Portugal and Spain were conducted over mobile telephones. Due to the relatively low fixed-line telephone coverage in Bulgaria, the Czech Republic, Estonia, Latvia, Lithuania, Hungary, Poland, Romania and Slovakia, 300 individuals were sampled and interviewed on a face-to-face basis. To correct for sampling disparities, a post-stratification weighting of the results was implemented, based on key socio-demographic variables[2].

The *Flash Eurobarometer* Intergenerational solidarity was conducted in order to examine EU citizens' opinions on:

- existing relations between the younger and older generations;
- costs of an ageing population – particularly in terms of pensions and elderly care;
- the need for pension and social security reforms;
- ways in which older people contribute to society – financially and more broadly;
- existing possibilities for autonomous living for elderly EU citizens;
- the provision of elderly care and support by social services;
- the role of public authorities in promoting intergenerational solidarity.

In among this vast quantity of data, the paper focussed on a few questions that were more specifically devoted to the social representation of elderly people (as a resource, rather than a social burden), considering how citizens' attitudes vary between countries and according to social categories (such as age, sex, education, urbanisation, occupation). The questions considered are:

- "Are older people are a burden for society?"
- "Are the media exaggerating the risk of a conflict between generations?"
- "In [our country], are there sufficient social services to support frail older people so that they can stay in their own home?"
- "Are people who have to care for older family members at home receiving good support from social services in [our country]?"
- "In the coming decades, will governments no longer be able to pay for pensions and care for older people?"
- "Is the financial help of parents and grandparents important for young adults who establish their own households and families?"
- "Do older people make a major contribution as volunteers in

[2] See *http://ec.europa.eu/public_opinion/flash/fl_269_en.pdf*.

charitable and community organisations in [our country]?"
- "Is the contribution of older people who care for family or other relatives not appreciated enough in [our country]?"

With this information it is possible to evaluate how elderly people are regarded by public opinion, how public services can support independent living for elderly people and family care-givers, how sustainable people feel an ageing society is, and the extent to which elderly people support younger generations in family and social life[3].

a) Older people are a burden for society.

In all Member States, at least two-thirds of EU citizens *somewhat* or *strongly disagreed* that older people are a burden on society: the total level of disagreement ranged from 66% in Lithuania to 95% in the Netherlands. Furthermore, a majority of respondents in 19 Member States, and a relative majority in a further eight, *strongly disagreed* that older people are a burden on society. Respondents in Cyprus and Greece were the most likely to *strongly disagree* (together with Ireland, while those in the Czech Republic were the least likely to do so (82% and 81%, respectively, vs. 37%). This more negative attitude towards the elderly seems to be perceived more strongly in the Eastern European countries (Bulgaria, Latvia, Hungary, Lithuania, Slovenia and Slovakia), but also in Portugal and in Malta.

Figure 1: Older people are a burden on society

Q1. I am going to read out a number of statements about relations between younger and older people. For each one, please tell me if you strongly agree, somewhat agree, somewhat disagree or strongly disagree
Base: all respondents, % by country

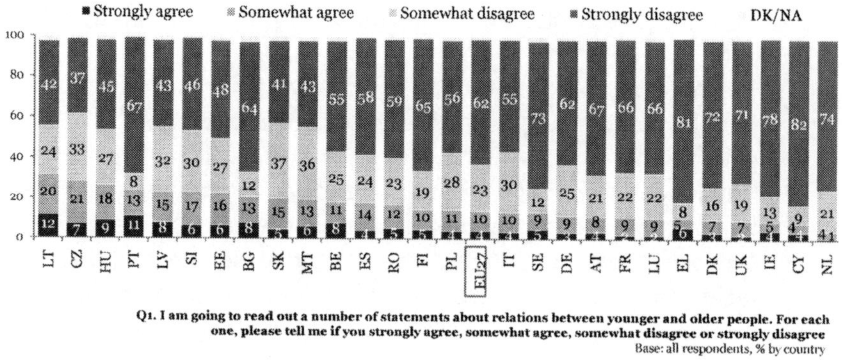

Source: *Flash Eurobarometer 269 (2009).*

Younger respondents did not necessarily see *older people as a burden on society*; the oldest respondents (over 64) and retirees were the most likely

---
[3] Data description is mostly quoted from the report.

to agree with this statement (25% and 22%, respectively, compared to, for example, 12% of 15-24 year-olds and 16% of 55-64 year-olds).

b) The media are exaggerating the risk of a conflict between generations.

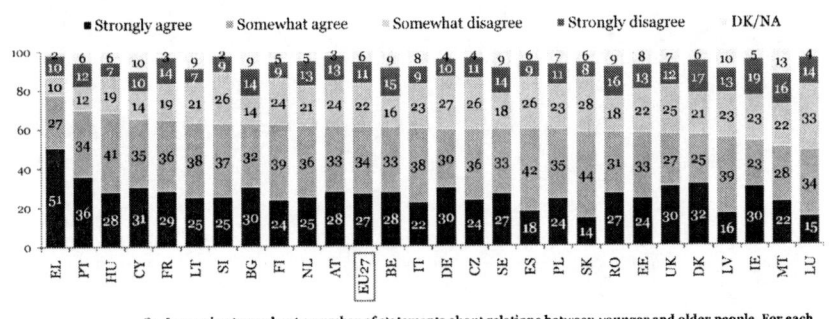

Slightly more than 6 in 10 EU citizens thought that *the media exaggerates the risk of a conflict between generations*: 27% *strongly agreed* and 34% *somewhat agreed* with this proposition.

Greek and Portuguese respondents were also the most likely to think that the media exaggerates the risk of a conflict between generations: 78% of Greek and 70% of Portuguese respondents *somewhat* or *strongly agreed* that this is the case. Although the total level of agreement was rather similar in Hungary and Portugal (69% and 70% respectively), only 28% of Hungarians *strongly agreed* that the media exaggerates the risk of a conflict between generations – compared to 36% of Portuguese respondents. Greek respondents were once again the most likely to *strongly agree* with this proposition (51%).

In Luxembourg, on the other hand, only 49% of respondents *somewhat* or *strongly agreed* - and a similar proportion (47%) disagreed - that the media exaggerates the risk of a conflict between generations. In all other countries (except Ireland), less than 4 in 10 respondents somewhat or *strongly disagreed* that this is the case, and the proportion ranged from 20% in Greece to 38% in Malta and Denmark. In Ireland, in total, 42% of interviewees disagreed with this statement.

Figure 2: The media are exaggerating the risk of a conflict between generations

Q1. I am going to read out a number of statements about relations between younger and older people. For each one, please tell me if you strongly agree, somewhat agree, somewhat disagree or strongly disagree
Base: all respondents, % by country

Source: *Flash Eurobarometer 269 (2009).*

c) *In [our country], there are sufficient social services to support frail older people so that they can stay in their own home.*

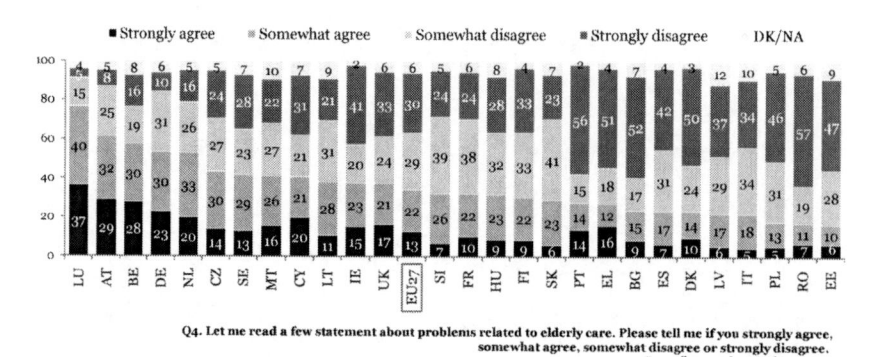

Only slightly more than a third of EU citizens in total *agreed* - and 59% *disagreed* - that there are sufficient social services in their country to support frail older people so that they can stay living in their own home.

Respondents in Luxembourg were the most likely to feel that *there are sufficient social services in Luxembourg to allow frail older people to stay in their own homes:* 37% of Luxembourgers *strongly agreed* and 40% *somewhat agreed.* In four other countries, a slim majority, at least, *somewhat* or *strongly agreed* with this statement: Austria (61%), Belgium (58%), Germany and the Netherlands (both 53%). In Estonia, Romania and Poland, on the other hand, at least three-quarters of interviewees *disagreed* that there are sufficient social services to support frail older people so that they can stay in their own homes (between 75% and 77% *strongly* and *somewhat disagree* responses). Furthermore, almost half of Estonians (47%) and Poles (46%) *strongly disagreed* that this was the situation in their country; in Romania, almost 6 in 10 (57%) interviewees *strongly disagreed.* Other countries where at least half of interviewees *strongly disagreed* were: Portugal (56%), Bulgaria (52%), Greece (51%) and Denmark (50%).

Figure 3: In [our country], there are sufficient social services to support frail older people so that they can stay in their own home.

Source: *Flash Eurobarometer 269 (2009).*

Both the youngest (under 25) and the oldest respondents (over 64) were more likely than respondents in the other age categories to think that there are *sufficient social services in their country to support frail older people so that that they can stay living in their own home.* In accordance with the above findings, it was also noted that full-time students, retired respondents and those with the lowest level of education were more likely to agree that there was sufficient support from social services. For example, while 40% of retirees agreed that there are sufficient social services in their country to support frail older people so that they can stay in their own home, roughly only a third of respondents in the other occupational groups agreed that this is the case: 32% of employees, 33% of "other" non-working respondents, 34% of self-employed respondents and 35% of manual workers.

d) People who have to care for older family members at home receive good support from social services in [our country].

Two-thirds of interviewees *disagreed* that people with caring responsibilities for older family members at home receive good support from their country's social services (35% *strongly disagreed* and 30% *somewhat disagreed*).

Similar to results obtained for the EU overall, respondents in almost all Member States were even less likely to agree that *people who have a responsibility of care for older family members at home receive good support from social services* than they were to agree to that there are sufficient social services for elderly people living on their own. In only one country - Luxembourg (54%) - did more than half of respondents *somewhat* or *strongly agree* that there is enough support for family members with caring responsibilities for older family members, while in more than half of the EU Member States more than 6 in 10 respondents *disagreed* that this is the case.

Focusing on those respondents choosing the more extreme negative response - i.e. *strongly disagree* - it was noted that while only a minority (7%) of Luxembourgish respondents chose this possibility, in Portugal, Bulgaria and Greece approximately 6 in 10 respondents *strongly disagreed* (between 57% and 64%). Respondents in the latter group of countries were not only among the most dissatisfied with support from social services for elderly people living on their own (as seen above), they were also the most dissatisfied with social services support for individuals who have a responsibility of care for older family members at home. Finally, a significant number of respondents in most Member States found it difficult to answer this question; the proportion of *don't know* responses ranged from roughly 1 in 20 respondents in Portugal, Finland, Ireland, Spain and Greece to at least one-sixth in Latvia (21%), Luxembourg (19%) and Malta (18%).

Figure 4: People who have to care for older family members at home receive good support from social services in [our country].

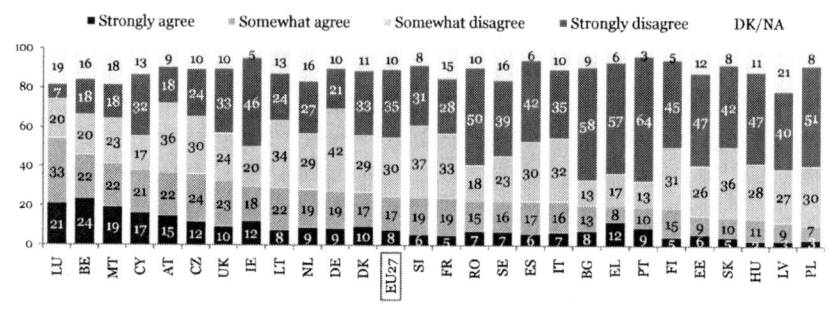

[Stacked bar chart showing percentages by country (LU, BE, MT, CY, AT, CZ, UK, IE, LT, NL, DE, DK, EU27, SI, FR, RO, SE, ES, IT, BG, EL, PT, FI, EE, SK, HU, LV, PL) with categories: Strongly agree, Somewhat agree, Somewhat disagree, Strongly disagree, DK/NA]

Q4. Let me read a few statement about problems related to elderly care. Please tell me if you strongly agree, somewhat agree, somewhat disagree or strongly disagree.
Base: all respondents, % by country

Source: *Flash Eurobarometer 269 (2009).*

Both the youngest (under 25) and the oldest respondents (over 64) were more likely than other age categories to think that *people caring for older family members at home receive good support from social services* in their country. For example, while 28% of those over 64 and 30% of 15-24 year-olds agreed that people caring for older family members at home receive good support from social services in their country, only between 22% and 24% in the other age categories agreed with this statement.

e) In the coming decades, governments will no longer be able to pay for pensions and care for older people.

Almost 6 in 10 respondents recognised that, in the coming decades, governments will no longer be able to pay for pensions and elderly care (25% *strongly agreed* and 33% *somewhat agreed*).

The statement received a total level of agreement ranging from approximately 4 in 10 interviewees in Bulgaria and Romania (38% and 40%, respectively) to twice as many in Portugal (81%). Other countries at the higher end of the distribution - with more than two-thirds of interviewees doubting about the affordability of pensions and elderly care - were Germany (72% *somewhat* or *strongly agreed*) and Austria (68%). Portuguese respondents were also the most likely to *strongly agree* with this proposition (54%), followed by Greek and German respondents (41% and 38%, respectively). In all other countries, not more than 3 in 10 respondents *strongly agreed*. Focusing on those choosing the more extreme negative response - i.e. *strongly disagree* - it was noted that less than 1 in 10 Germans, Slovaks, Czechs and Italians chose this possibility, while in Romania and Bulgaria the proportion was more than three times higher (31% in both countries).

Figure 5: In the coming decades, governments will no longer be able to pay for pensions and care for older people.

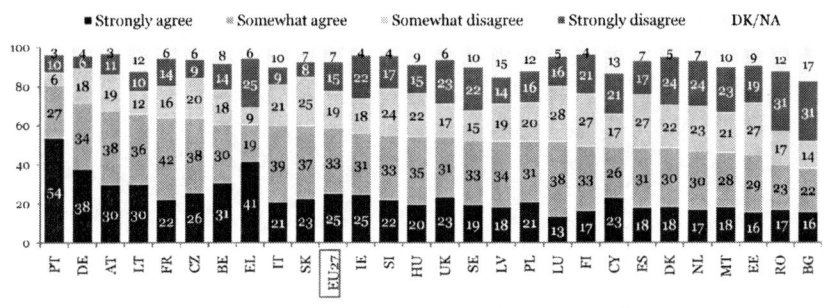

Q2. Now I would read out a few statements related to pensions. Please tell me if you strongly agree, somewhat agree, somewhat disagree or strongly disagree.
Base: all respondents, % by country

Source: *Flash Eurobarometer 269 (2009)*.

Respondents aged between 25 and 54, those with higher levels of education and a higher occupational status were the most concerned about the affordability of pensions: roughly 6 in 10 of these respondents somewhat or *strongly agreed* that, *in coming decades, governments will no longer be able to pay for pensions and care for older people*, compared to, for example, a slim majority of retirees or respondents with the lowest level of education (both 53%).

f) The financial help of parents and grandparents is important for young adults who establish their own households and families.

In total, almost 9 in 10 EU citizens agreed - and a slim majority (55%) *strongly agreed* - that financial help from parents and grandparents is important when young adults are starting to establish their own households and families.

In almost all Member States, there was almost no doubt that *financial help from parents and grandparents is important for young adults establishing their own households and families:* more than 8 in 10 respondents in 23 Member States somewhat or *strongly agreed* with this statement. The total level of agreement, however, was considerably lower in Denmark (59%), the Netherlands (65%), the Czech Republic (71%) and Sweden (76%). Furthermore, while at least 8 in 10 Portuguese, Greek and Cypriot interviewees *strongly agreed* that parents' and grandparents' financial help is important for young adults forming their own households and families, only half as many, or fewer, interviewees in the last-named countries - and Slovakia - *strongly agreed* that such financial support is important (29% in the Netherlands and Denmark and between 38% and 40% in the Czech Republic, Slovakia and Sweden).

Figure 6: The financial help of parents and grandparents is important for young adults who establish their own households and families.

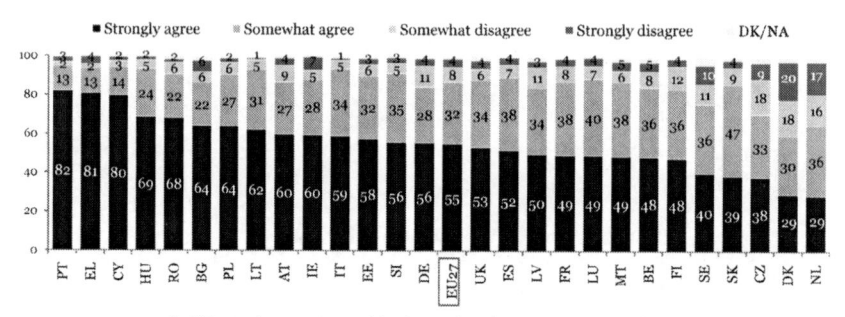

[Bar chart showing survey responses by country with categories: Strongly agree, Somewhat agree, Somewhat disagree, Strongly disagree, DK/NA. Countries from left to right: PT, EL, CY, HU, RO, BG, PL, LT, AT, IE, IT, EE, SI, DE, EU27, UK, ES, LV, FR, LU, MT, BE, FI, SE, SK, CZ, DK, NL.]

Q3. Older people are not just receiving from society, they can also give something back. Please tell me if you strongly agree, somewhat agree, somewhat disagree or strongly disagree.
Base: all respondents, % by country

Source: *Flash Eurobarometer 269 (2009).*

The results for the statement that *financial help from parents and grandparents is important when young adults are establishing their own households and families* showed significantly less variation across socio-demographic groups. It did appear, however, that the over 54 year-olds and retired respondents were more likely than their counterparts to express strong agreement (58-60% compared to, for example, 51% of 15-24 year-olds and 54% of 25-39 year-olds).

g) Older people make a major contribution as volunteers in charitable and community organisations in [our country].

A large majority of EU citizens also agreed that older people make a major contribution to society via voluntary work in charitable and community organisations in their country (44% *strongly agreed* and 34% *somewhat agreed*). The total level of agreement with this statement ranged from around 4 in 10 respondents in Poland (39%) and Romania (43%) to more than 9 in 10 of the Irish, British, Portuguese and Dutch interviewees (between 91% and 95%).

The eight Member States where respondents were the least likely to agree with this all belong to the group of countries that joined the EU in 2004 or later; the eight countries where respondents most frequently agreed were all pre-2004 enlargement countries. In almost all countries of the latter group, at least half of respondents *strongly agreed* - and less than one-tenth *somewhat* or *strongly disagreed* - that older people's voluntary work contributes to society in important ways. Portuguese, Irish and British respondents were the most likely to *strongly agree* with the statement (71%, 69% and 65%, respectively). In the former group of countries (except for Latvia), only between 15% and 26% *strongly agreed* that there is a major contribution from older people performing voluntary work, while between 27% and 47% *somewhat* or *strongly disagreed* that this is the case. Romanian respondents were the most likely to *strongly disagree* (25%), followed by Bulgarian and Czech respondents (18% and 17%, respectively). In Latvia, however, only 18% in total disagreed with the statement and 21% provided a don't know response.

Figure 7: Older people make a major contribution as volunteers in charitable and community organisations in [our country].

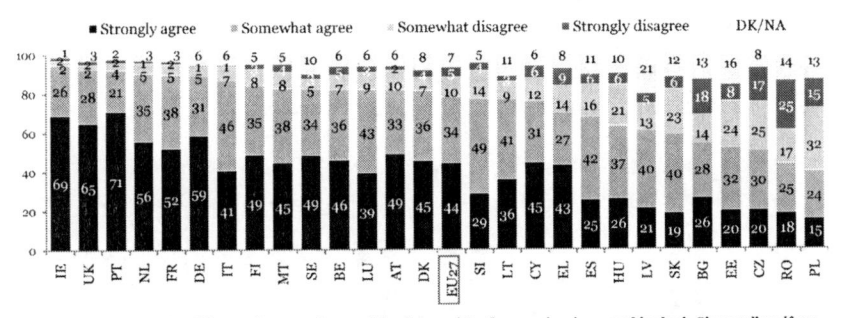

Q3. Older people are not just receiving from society, they can also give something back. Please tell me if you strongly agree, somewhat agree, somewhat disagree or strongly disagree.

Source: *Flash Eurobarometer 269 (2009).*

Only 7 in 10 of the 15-24 year-olds and full-time students *somewhat* or *strongly agreed* that older people make a major contribution to society via voluntary work in charitable and community organisations in their country. The total level of agreement increased to more than 80% for the over 54 year-olds, retirees and those with the lowest level of education. Rural residents were more likely than city dwellers to *somewhat* or *strongly agree* that older people's voluntary work makes an important contribution to society (80% vs. 74% in metropolitan areas). Finally, men and women held relatively similar views about the contribution of older people to society.

h) The contribution of older people who care for family or other relatives is not appreciated enough in [our country].

Slightly more than three-quarters of interviewees thought that the contribution of older people who care for family members or relatives is not sufficiently appreciated in their country (44% *strongly agreed* and 33% *somewhat agreed*).
 Respondents in Portugal (91%), the UK (87%) and Finland (85%) were the most apt to *somewhat* or *strongly agree* with this proposition, while respondents in Luxembourg were the least likely to do so (58%). Luxembourg

was the only country where more than 3 in 10 (32%) respondents *somewhat* or *strongly disagreed* that older people's contribution in this respect was not being sufficiently appreciated. Portuguese respondents stood out from the pack *somewhat* as roughly three-quarters (74%) *strongly agreed* that the contribution to society by older people, who have a responsibility of care for family members or relatives, is not appreciated enough in their country. In Germany, Finland, Bulgaria, Ireland and the UK, between 5 and 6 in 10 respondents expressed their strong agreement, while in Luxembourg, Estonia, France, Italy, Malta, Slovenia, Lithuania and Slovakia, only between a quarter and a third *strongly agreed*.

Figure 8: The contribution of older people who care for family or other relatives is not appreciated enough in [our country].

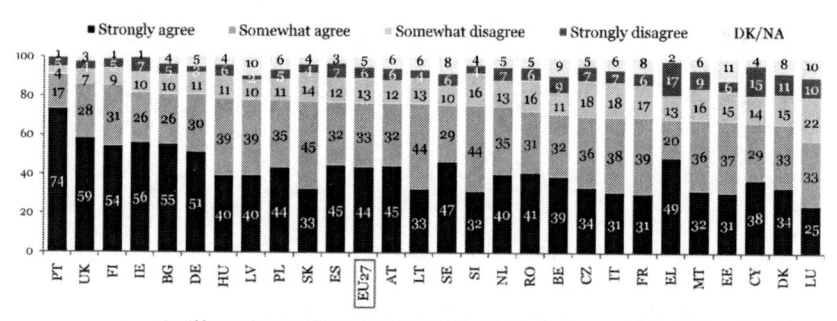

Source: *Flash Eurobarometer 269 (2009)*.

Respondents between 40 and 64 years of age were the most likely to *somewhat* or *strongly agree* that the contribution of older people who have a responsibility of care for family members or relatives is not sufficiently appreciated in their country, while 15-24 year-olds (and full-time students) were the least likely to do so (80% vs. 71%-72%). The results by occupational status showed that employees were the most likely to *somewhat* or *strongly agree* with the above statement (80% compared to 74-76% in the other occupational groups); however, when looking at those who *strongly agreed* with the statement, it appears that retirees were just as likely as employees to select this possibility (47% *strongly agreed* vs. 46% of employees).

4. Final remarks

According to the perceptions of the majority of EU citizens, the social representation of elderly people and their role in intergenerational solidarity is rather good:

- Elderly people are not considered a burden for society by two-thirds of respondents, and about 60% of people believe that media are exaggerating the risk of an intergenerational conflict.
- conversely, the positive role of public intervention supporting Elderly people is not so strongly shared by respondents: only one-third of interviewees believes that social services are sufficient to maintain frail older people at home or believe that people who care for elderly relatives are adequately supported by them; only about 40% of respondents believe that in the future governments will be able to pay for pensions and care for elderly people; not surprisingly, differences among countries here are very high.
- The vast majority of respondents believe that elderly people are a very important resource for other generations, in family relations and in society; almost 90% believe that parents and grandparents are financially helping generations to set up new families, and almost 80% believe that the voluntary work of elderly people in society is very important.
- Finally, and rather controversially, more than 70% of respondents believe that the contribution of elderly people in family relations is under-appreciated.

Moreover, national responses vary significantly from one country to another (in some questions the distance between the higher and the lower percentage is more than 50%), while fewer differences can be found according to sex, working conditions, urbanisation; only age seems to cause slight differences in attitudes, but sometimes elderly people and younger generations give similar responses, while adults significantly differ. In other words, it can be said that variations in opinions appear to be more determined by the general social environment (cultural, social welfare systems) than by individual personal condition (including a possible corporative plea for the protection of a single generation against other generations' interests).

European citizens seem to clearly acknowledge the existence and the importance of intergenerational solidarity, inside family relationships but also in social life (through voluntary work and public redistribution of resources and services by the welfare state). Policy makers therefore have to carefully consider the intergenerational dimension of social and family policies, promoting the existing reciprocal exchange of resources inside family networks, and shaping their national and local policies and services in an intergenerational relational approach. This could be a powerful tool for coping in a positive way with the current demographic transition, supporting both younger and older generations, thereby preventing a possible - and dangerous - social and economic conflict between generations. This is potentially one of the most innovative approaches that European welfare systems could adopt in the twenty-first century.

References

All graphs from *Flash Eurobarometer 269* used with permission. Source: http://ec.europa.eu/public_opinion/flash/fl_269_en.pdf. The European Union does not endorse changes, if any, made to the original data and, in general terms to the original survey, and such changes are the sole responsibility of the author and not the EU.

- Bramanti, D. (ed.) (2001). *La famiglia tra le generazioni [The family across the generations]*. Proceedings of the Conference of the Centre for Family Research and Studies, Catholic University, Milan, 13-14 October 2000, Vita E Pensiero, Milan.
- Donati, P. (1991). *Secondo Rapporto sulla famiglia in Italia [Second Report on the Family in Italy]*. Edizioni Paoline, Cinisello B., 11-108.
- Donati, P. (1991). *Quarto Rapporto sulla famiglia in Italia (Fourth Report on the Family in Italy)*. Edizioni San Paolo, Cinisello B., 11-87, 383-404.
- Donati, P. (2007). *Ri-conoscere la famiglia: quale valore aggiunto per la persona e la società? Decimo Rapporto Cisf sulla famiglia in Italia [To newly acknowledge the family: what added value for persons and for society?]*, Edizioni San Paolo, Cinisello B., 13-62, 385-412.
- Facchini, C. (ed.) (2008). *Conti aperti. Denaro, asimmetrie di coppia e solidarietà tra le generazioni. [Open accounts. Money, couple asymmetries and intergenerational solidarity]*. Bologna: il Mulino.
- Fors, S. & Lennartson, C. (2008). *Social Mobility, Geographical Proximity and Intergenerational Family Contact in Sweden, Ageing and Society* 2: 253-270.
- Green Paper *"Confronting demographic change: a new solidarity between the generations"*, Communication from the Commission, Brussels, 16.3.2005.
- Hoff, A. (2007). *Patterns of Intergenerational Support in Grandparent-Grandchild and Parent-Child Relationships in Germany, Ageing and Society* 5: 643-666.
- Luescher, K. (2000). *Intergenerational Solidarity or Intergenerational Ambivalence?*, Family Observer 3: 28-31.
- Monserud, M. (2008). *Intergenerational Relationships and Affectual Solidarity Between Grandparents and Young Adults*. Journal of Marriage and the Family 1: 182-195.
- *Promoting Solidarity between the Generations*. Communication from the Commission to the European Parliament, the Council, the European Economic and Social Committee and the Committee of the Regions, Brussels, 10.05.2007

2.5 Intergenerational Solidarity: Rebuilding the Texture of Cities

Lorenza Rebuzzini
Forum delle Associazioni Familiari

Everybody experiences in his/her life intergenerational relationships and solidarity, but only in recent years has it become a policy issue, particularly in urban areas. The mix of people isolation and ageing has become a critical point, especially where the young and the old are competing for resources, public space and attention.

As previously noted:

> "This has been further exacerbated by the way policies and services are normally developed around targeted groups or issues that are by their nature disjointed and discriminatory. The aim of intergenerational work is to find ways to develop and strengthen these relationships and consequently become an agent of social change with benefits to the whole of society"
> (Municipality of Manchester, Looking Backward, Looking Forward).

Combating isolation, rebuilding the social texture, rebuilding "good neighbour" relationships, and promoting active ageing practices is today a necessity felt in many cities, especially those in which an ageing population, migration, and high levels of unemployment among young people are mixed together. Therefore, policies to enhance intergenerational solidarity are strictly linked to the wellbeing of society, and therefore to the wellbeing of families who are a part of society.

In this article two good practices, two projects developed in Turin (Italy) and Manchester (UK), will be analysed, in order to understand which characteristics in intergenerational policies make them effective. Both Turin and Manchester are mid-sized cities with large metropolitan/suburban areas (Turin has 865,263 inhabitants, Manchester 464,200, but if we take into consideration the whole Metropolitan area, they both have almost 2,000,000 residents). They both developed as industrial cities, although at different times, and their industries underwent severe crises; they have both been affected by processes of internal and external migration, ageing of the migrant population, and high levels of unemployment among young and disadvantaged people. In addition, in the nineties they both experienced strong urban regeneration programmes and commercial revitalisation, changing their productive and economic assets.

The contents of this article are based on interviews with Renato Bergamin (Director of Project Cascina Roccafranca, Turin), and Paul McGarry (Senior Strategy Manager, *Manchester Generations Together* programme). Both projects have been developed by the municipalities of Turin and Manchester and are based on strong alliances with local stakeholders (foundations, voluntary or family associations, universities), although they have been built in two different (sometimes opposite) ways and according to different goals: while the *Manchester Generations Together* programme specifically targets intergenerational solidarity, Cascina Roccafranca has been developed as an urban development plan. Nevertheless, in both cases, the outcome was an intergenerational approach to tackling isolation and rebuilding the urban fabric.

Manchester Generations Together

Manchester Generations Together is a programme started in 2009, with funding due to end at the end of March 2011 (though the municipality of Manchester is putting in place plans to continue it). This programme is part of a larger project, the *Valuing Older People* (VOP) Project, launched in 2003 by Manchester City Council, the three Manchester Primary Care Trusts and community and voluntary organisations. The partners' aim was to improve the quality of life of Manchester's older adults by working together.

Valuing Older People represents a commitment to improve services and opportunities for the city's older population. It also challenges Manchester's public agencies, businesses and communities to place older people at the centre of the extensive plans for the regeneration and reshaping of the city. VOP soon developed an interest in Intergenerational Practice (IP), and in 2006 started its close collaboration with the Beth Johnson Foundation, a UK charity, which convenes the UK Centre for Intergenerational Practice. The first phase of the VOP Project, before Generations Together, was based on a report commissioned from the Beth Johnson Foundation, Looking Backward, Looking Forward, which included the following elements:

- training in intergenerational practice for over 100 front-line staff;
- stakeholder interviews and analysis;
- funding for a small number of demonstrator projects;
- establishment of a learning network;
- an Intergenerational Practice e-bulletin;
- strategy and policy development (how 'intergenerational practice' adds value to Manchester City Council's corporate, departmental and partnership priorities and how it will improve the lives of residents).

One of the strongest messages that came through from the report was that there was a real need to establish opportunities to connect people across the generations to build understanding and respect. A large proportion of interviewees described age segmentation as an increasing part of our society, manifesting itself in decreased contact between younger and older people. In looking at community cohesion, it is important to begin to explore and understand the different world views of the different generations. It is also necessary to acknowledge that tensions between generations are not a new phenomenon, as each new generation strives to develop its own identity and place in society. Indeed, it is the way these relationships are negotiated and established that is key.

It is also important to recognise the role of the extended family where this still exists. In this respect, the use of storytelling techniques of group learning across ethnic groups has proved to be effective. In the Netherlands, for example, a programme called *A Neighbourhood Full of Stories* has been developed. The Netherlands Institute of Care and Welfare (NIZW) has developed a new method for promoting the integration of generations and cultures: 'neighbourhood-reminiscence'. This method uses memories and stories of neighbourhood residents in order to promote exchanges, mutual understanding and respect between different age and cultural groups. Neighbourhood-reminiscence is a community development, based on the local neighbourhood level (Mercken, 2003).

The "Intergenerational Programme is therefore about building generational relationships within community settings between people. It is also a way of addressing social exclusion of older and younger people and making places friendly for people of all ages", says Paul McGarry; "Intergenerational approaches are an effective way to address a number of issues - many of them key government priorities - such as building active communities, promoting citizenship, regenerating neighbourhoods and addressing inequality and social exclusion".

Interest in the IP has developed in the context of a number of social policy concerns often linked to community cohesion and social exclusion in disadvantaged neighbourhoods. This has included concerns about levels of anti-social behaviour and joblessness, in particular involving young men, and addressing issues that affect older people, such as loneliness and depression: "This has led us to seeing IP as a tool for improving the quality of life for older and younger people in Manchester and informed our work developing Manchester as an 'Age Friendly City'. We have recently been accepted into the WHO Age Friendly City network"[1].

[1] See *http://www.who.int/mediacentre/news/releases/2010/age_friendly_cities_20100628/en/index.html*.

The Valuing Older People team co-ordinates Manchester's *Generation Together* programme. The team is located in the Manchester Joint Health Unit, a public health team based in Manchester City Council, co-ordinated by a Programme Manager (Patrick Hanfling), a Programme Officer (Rachel York) a Community Engagement Officer (Tracey Annette) and the leader of the IP demonstrator work (Programme Manager Sally Chandler). Progress is reported to the Senior Strategy Manager (Paul McGarry) and a wider Steering Group of senior managers in the Council. The programme is therefore delivered through Manchester City Council, the voluntary and community sector and the academic sector, through Manchester Metropolitan University and the Manchester School of Architecture.

Manchester Generation Together was funded by the previous Labour Government: £5.5 million was allocated for the programme, which local authorities could apply to (up to £400,000 each). Nearly all Local Authorities (132) in England applied and Manchester was one of the 12 successful bids. Manchester's bid involves 13 projects based around four themes: shared spaces, shared skills and learning, health and wellbeing, families.

Five of the Manchester projects will be run through Manchester City Council, six through the voluntary sector and two by academic bodies.

1. **ALL FM community radio project** (in the district of Levenshulme), is built on the results of past projects, which challenged the negative perceptions that different generations have of each other. This project will target multi-cultural neighbourhoods to identify older and younger volunteers to learn all aspects of broadcasting.
2. **Food Futures Cookery Classes** (city-wide) involves young people not in employment, education or training and isolated older people producing healthy meals together.
3. **Manchester City Council Youth Service's** Intergenerational Volunteers in Schools develops sustainable school volunteering programmes, involving grandparents and parents in skill sharing.
4. **Manchester School of Architecture**, architectures of intergenerational engagement, raises awareness of the design implications of shared spaces.
5. **Home Improvement Agencies' maintenance skills exchange** involves Do-It-Yourself (DIY[2]) skills taster days, DIY training and a makeover of a community building by an intergenerational team.

---

[2] "a term used to describe building, modifying, or repairing of something without the aid of experts or professionals" *(https://secure.wikimedia.org/wikipedia/en/wiki/DIY).*

6. **Manchester Adult Education Service Adult Education Intergenerational Buddy Exchange** uses Adult Learner volunteers to help vulnerable families by offering support to young mothers and learning support to children and young people.
7. **Generation Games** involves extended families in games and interactive activities to facilitate better communication between family members. It helps them to develop mechanisms that boost families' capacity to support children while supporting the adults to become further engaged in volunteering, learning or employment.
8. The project between *Community Service Volunteers (CSV)* and the *Powerhouse Library*, **'Young and Older Voices,'** focuses on Moss Side, one of the most culturally diverse areas in the city. The project develops more hands-on intergenerational volunteering opportunities that sees older and younger people becoming active citizens and advocates for social change.
9. **The Multicultural Cookbook & Community Allotment** and **Inspired Sisters** projects provide opportunities for children and young people to learn about food growing and sustainable living, develop cooking skills and experience preparing and sharing food from other cultures. All participants benefit from physical activity on allotments.
10. **The Roby Mental Health Project** aims to equip groups of older and younger people with advocacy and advice-giving skills as tools to address mental health issues within their communities.
11. **Common Ground** involves a Big Brother-style café conversation, which teases out attitudes towards people from different generations. Participants then work together on shared tasks and the process is recorded through video diaries to monitor and record changing attitudes as bridges are built.
12. **Intergenerational Evaluators** involves training up younger and older people to be able to carry out evaluation of intergenerational projects and programmes. This projects aims at enabling people to work together and start social and non-profit enterprises.

The last project has been developed by a gay and lesbian association. All of these projects will be evaluated and monitored by a specially appointed national agency. Moreover, an independent research organisation called York Consulting has been appointed to evaluate the Generations Together programme. York Consulting has developed different approaches to collect information about each project, including an online management information tool and telephone interviews. Manchester is one of six local authorities to have been chosen as a case study site, where more in-depth evaluation of some projects will be done.

"Plans for the future involve the development of IP-influenced policy with our partners, cities and districts, a partnership with BJF and Leeds local authority to develop a toolkit on IP, a specific programme involving Age Friendly Cities, the development of new skills by community development workers, exploration of UK and European research opportunities in collaboration with partners", concludes Paul McGarry.

Cascina Roccafranca, "The Art of Building a Common Space"

Cascina Roccafranca was an abandoned farm-building in Mirafiori district, North of Turin, where a Fiat factory was established in 1939 and is still operating. In 2007, thanks to funds granted by the EU project Urban 2, the farm building was converted into a cultural and recreational centre.

The conversion was carried out by a team of architects in strict collaboration with the team of social workers appointed to the project by the Municipality of Turin. "The multidisciplinary approach and the possibility of projecting together was shown to be an essential element in building a common space, a 'home for the district'", says Renato Bergamini. This was an uncommon and challenging, but very promising, approach to the conceptual work of converting the building in order to create 'a place for the district'. Therefore, from the very beginning, Cascina Roccafranca was meant to become an *aesthetically valuable* and at the same time a *functional place*: large windows, empty but cosy and interchangeable spaces in which many different activities can be developed and many different age groups can meet.

Cascina Roccafranca is run and managed by a Foundation of which the Municipality of Turin is the Founder, while associations and organisations from the district (parishes, schools, informal groups) are the Main Partners. The Executive Board is made up of three persons nominated by the Mayor, and three persons representing district associations, including the "Gruppo Abele" association[3], which plays a prominent role. The Foundation has five main goals: to build citizenship, to enhance the wellbeing of the community, to promote a mainstreaming culture based on solidarity and linkage to the territory, to conduct an experiment in social partnership between public and private sector, and to promote a culture of respect for different people. "The two keywords for understanding the project", says Riccardo Bergamini, "are 'Welcoming' and 'Participation'".

[3] See http://www.gruppoabele.it.

Cascina Roccafranca is intended to be a free and welcoming space where the staff shares and promotes a bottom-up approach, according to the following guidelines:

- **Creating synergies among stakeholders:** "Experience has taught us that content is important, but not fundamental. Instead, your methodology is essential and must involve the participation of stakeholders, informal groups, and local associations. Social workers have to allow space for interests brought by the people, even if they think such content is irrelevant. The social workers' point of view is one point of view, it is not the point of view. We propose a dialogic method in planning events, actions and initiatives". According to this methodology, Cascina Roccafranca's staff try to create synergies among the different stakeholders that share similar interests, ideas and projects.
- **Using a plurality of languages:** "Many different forms of communication are used to convey messages: music, dance, theatre, etc. The storytelling technique has also been developed in an intergenerational project, I Nonni raccontano ('grandparents telling stories'), in which older people share their memories with younger people on how the district was in past times.
- **Increasing skills:** "Groups, associations, persons who come to Cascina Roccafranca have one need above all: to be listened to", says Riccardo Bergamin. In Cascina Roccafranca calls are launched to fund and support micro-projecting and working groups. Riccardo Bergamini notices that "in these years we have realised that the management of Cascina Roccafranca, in terms of schedules, deadlines, communication, is a complex and necessary job, nevertheless we have realised that it is much more necessary to have a greater capacity to listen to the ideas and interests of people, and to enhance skills that are already present. Acknowledging the role and the skills of others, especially when they are non-professional, is a difficult but necessary step".
- **Building formal and informal networks:** The Cascina Roccafranca staff try to promote formal and informal networks based on common interests, maintaining them open to new people who want to join and bring new ideas.

In this very de-structured and open approach, the encounter between different age groups is left to the freedom of people getting together in this space and building formal or informal networks, based on common interests and projects. The aim is therefore to build the setting where intergenerational solidarity can be developed. Services and activities consistently delivered at Cascina Roccafranca are:

1. **Info Point:** information on all activities in the centre and the district.
2. **Counselling Services:** legal advice, trade union services, information on the wellbeing of the over-sixties, support for foster and adoptive families, counselling for parents, help for victims of violence, information and counselling for those who are ill and their right to be assisted, information for owners of animals; all of these counselling services are run in collaboration with voluntary associations or groups.
3. **Restaurant La Piola dell'Incontro** and **Café ¿Algomas?:** organic and fair trade products and products from local micro-breweries.
4. **Centre for families "The Enchanted Fortress":** with play activities and parental counselling.
5. **The Ecomuseum:** with a large section dedicated to the history and development of Mirafiori District.
6. **Play, Move, Become Friends:** dedicated to children and families, run by Agape Foundation, on Saturday mornings and Sundays.
7. **Critical Consumers' Shop:** organised by the group of 150 families that have associated in the Cascina Roccafranca Solidal Purchasing Group. The Project is meant to spread new lifestyles and pays special attention to the quality of life. The group also gives attention to the issue of responsible tourism and has opened an info point on it, inside the shop.
8. **Women's Space:** run by an informal group of women, focusing on the following themes: generational solidarity, work, health, history of women, culture and arts.
9. **Wellbeing Space:** run by a group of associations already interested in the theme; gym courses and conferences are organised.
10. **Cascina Together:** a project dedicated to people who have free time during the day (e.g. pensioners, stay-at-home mums, unemployed people, etc.). Activities and self-run courses are organised.

11. Incubator of Ideas: an activity in which proposals for new projects are gathered together, examined and promoted. A number of projects have been developed as a result – time bank, a social platform based on the internet[4], Roccafranca Film (cineforum), free software developing projects and organisation of a Linux Day, activities based on intercultural exchange with Arab migrants.

12. Cultural Events; each month, cultural events are organised.

This list shows that there is "room for every generation". Nevertheless there is a missing age group: adolescents. This is partly due, according to Bergamin, to the fact that a centre for young people has just opened near Cascina Roccafranca. But it is also due to the specific age group, and the "impossibility" of adolescents participating in such a context. The presence of young adults' (20-35 years old) should be strengthened as well: this age is taking part in very specific projects (e.g. the creation of open source software), while a daily, more integrated and plural experience is still missing. The presence of families is massive, and this is of course a "natural place" in which intergenerational relations can be built.

Nevertheless, 30 per cent of the regulars going to Cascina Roccafranca are aged 60 or over. They represent the majority of the one hundred volunteers working there, with different levels of involvement and participation. There are almost forty volunteers permanently associated with the management of Cascina Roccafranca (scheduling, gardening, co-ordinating groups, managing activities) and they are involved in co-ordination and periodic meetings. Moreover, there are sixty volunteers linked to specific projects and activities, belonging to associations or informal networks. "The over-sixties," says Riccardo Bergamin, "are a great asset in our work. They have two precious characteristics: they have time, and they have skills. They have time, because they no longer have to work and don't have to look after elderly parents or small children (or, if they do, it's not a full-time and daily activity), and they have strong skills, acquired at work or indeed over their whole lives. They are proud and willing to share their skills, and we try to emphasise and enhance this attitude: for example a judge, now retired, is running a course in the history of music, open to anyone who is interested in this subject".

In three years of activity Cascina Roccafranca has therefore become an open space, respected by all the people in the district: since opening, the building has never been vandalised, and there have been no thefts. In this open laboratory, policies to build intergenerational solidarity have been put

---

[4] See *http://www.laperquisa.it*.

to the test and singles, as well as families, have been "got off the ground" with the explicit intention of revitalising the district. The mix of public and private actors is intended to stress shared responsibility and participation, but also to enhance the fact that this project has to be economically sustainable. Projects and activities must be self-financed for 30 per cent of the total cost. The evaluation of the first two years of activity has been done. Single activities have been evaluated according to two parameters: effectiveness of the project itself, and effectiveness in relation to the strategic objectives of Cascina Roccafranca.

Conclusions

Turin and Manchester have developed two very different projects as regards the target, the approach and the methodology followed. Manchester has developed an explicit and coherent set of policies on intergenerational practice, while the Municipality of Turin has developed micro-projects based on participation and solidarity. These two experiences can be considered complementary in showing how the local and the micro-level can be seen to be fundamental in building intergenerational solidarity policies, as each urban reality is distinctive in terms of the age composition of inhabitants.

Both projects show the great importance of re-thinking and rebuilding social policies, as well as urban contexts, from an intergenerational point of view: where the pact among generations is recognised and enabled, and spaces are planned and built from an intergenerational and participatory point of view, the wellbeing of families and communities is reinforced. In addition, thanks to this approach, families can be greatly helped in recognising their inner - intergenerational (in essence) - structure. Working at local level means also facing a transformation of the welfare system and being more in touch with the "living spaces" of families. This leads us to a further consideration: intergenerational solidarity can be generated and promoted in the family when the family is considered the basic and prominent cell of the society, and is therefore supported and promoted as such. This also means that the alliance between family and society is built in contexts in which mutual recognition and acknowledgment between generations is promoted and recognised.

Re-thinking social policies and urban development starting from the intergenerational approach with an open point of view will therefore be an interesting, and necessary, challenge for the future of social policies. As demonstrated in Existential Field 4b of the FAMILYPLAT-

FORM Report on *Major Trends Local Politics – Programmes and Best Practice Models*[5] with regard to family policies, local policies and the management of community development can be great opportunities for developing policies of intergenerational solidarity.

Internet references

- *http://www.manchester.gov.uk/generationstogether*
- *http://www.manchester.gov.uk/vop*
- *http://www.cascinaroccafranca.it*
- *http://www.familyplatform.eu/en/1-major-trends/reports/4b-local-politics-programmes-and-best-practice-models*

Other references

- Bergamin, R., Bianco, L., Ieluzzi, S., *L'arte di costruire lo spazio comune*, Animazione Sociale 67 (2009): 67-75.
- Mercken, C. (2003). *Neighborhood-reminiscence: Integrating generations and cultures in the Netherlands.* Journal of Intergenerational Relationships, 1.1: 81-94.
- Municipality of Manchester (2007). *Looking Backward, Looking Forward.* http://www.centreforip.org.uk/Libraries/Local/949/Docs/Manchester%20Strategy%20-%20VOP.pdf.

---

[5] Available from *http://www.familyplatform.eu/en/1-major-trends/reports/4b-local-politics-programmes-and-best-practice-models*.

2.6 Solidarity in Large European Families

Raul Sanchez
Institut d'Estudis Superiors de la Família

Solidarity between generations is a reality experienced by many European large families on a daily basis. But the European social model that we are building does not seem to be well adapted to this reality. Two true stories from Spain, my country, can be used as examples.

The first is about a young couple who got a "mini apartment" from the City Council in a populous neighbourhood far from the city centre. It was the only affordable housing they could get to start their family life. They were between temporary contracts and casual jobs, and these prevented them from obtaining a simple mortgage. After a year they had a beautiful baby and the following year, twins! Thirty-five square meters, a couple and three children. They asked the City Council to provide them with another house. It was impossible: they had had the contract for eight years and could not sell or rent out the tiny apartment. They had to wait six more years. Of course, they were outraged and, above all, desperate, because in addition - as they are a bit revolutionary - they would like to have another child.

The second example is the story of a large family which recently moved from Belgium to Spain. The father is Belgian, the mother is from Spain. They have six children. The Belgian government transferred 1,200 Euros to the parents every month without any means test. This is a universal child benefit in that country. But after a while they had to adjust to the Spanish social system. Result: zero Euros - yes, nothing! - from the Spanish Social Security and 175 Euros per month from the regional government: 1,025 Euros less per month than before. The mother had to leave her job to take care of their children and, of course, they are already looking for new employment in the south of France, a country far more generous to large families than Spain.

These are two real examples of an emerging social model which ignores, or even rejects families that wish to have and bring up something as human, intimate and necessary as children. Despite small steps in recent years, all experts agree - and experience shows - that the distance to a family-friendly society is still very great.

The overall European framework for family issues has been characterised for many years by a very low birth rate, always below replacement levels; by the progressive incorporation of women working outside the home, with related social and economic changes, and by a labour market oblivious to this phenomenon and not well adjusted to the family aspect of workers' lives.

A globalised and competitive market leads to longer years of study, and also generates highly volatile employment among young people, and greater geographical mobility for everybody, separating them from their closest family network. Furthermore, housing is very expensive. For these reasons people are delaying the age at which they marry and have their first child. We are experiencing a rapid increase in divorce and in the number of births outside marriage and abortions. In addition, there are very large differences in measures of public family support among the member countries of the European Union.

From this quick survey it seems clear that the family has become one of the most overlooked social structures in Europe, and we are beginning to suffer the consequences. We have forgotten that the family is the largest NGO, the one that takes the best care of the sick, the elderly, and the unemployed, especially in times of crisis. When the family network is missing, social costs increase significantly. We have ignored the supportive role that families with children play in maintaining the celebrated European welfare system, and as a provider of human capital. That is why the focus is now on how to reduce pensions, on how to extend working life, and on how to introduce co-payments in health, social or educational services. In short, we are now reducing the quality of life of a system which has been kept up till now thanks to the brave men and women who have chosen to have kids in a society that is taking them less and less into consideration.

With all of this in mind, it seems increasingly difficult to have children, and especially difficult to have several of them. How have we been able to ignore such an important factor for the maintenance of the envied European social model?

Each survey held in European countries shows that women want to have more children than they are currently having[1]. As women's' responses in these surveys show, it would be enough to help families to carry out their own functions; supporting them so they can have as many children as they want, and giving them the opportunities and the necessary time to care and educate the children well.

To achieve this goal it is vital to avoid all kinds of social and economic penalties or discriminations due to family size, and support them with a series of allowances, either monetary or in services. This should be supported by governments at all levels, especially with a "courageous budget", devoting considerable resources to the long term, and viewing them as an investment, not an expenditure, since the future benefits are obvious to everyone.

---

[1] See *Eurobarometer 2006* on fertility and childbearing preferences: *http://ec.europa.eu/public_opinion/archives/ebs/ebs_253_en.pdf*.

It is not enough to invest money. A new mentality is required, to promote a social contract for the family - and for the children as a social good - in every country and at the European level, involving not only politicians but also the economic, educational, cultural and mass media world. This could be the beginning of a real family-friendly society, and the beginning of a truly European intergenerational solidarity.

Without families and children there is no welfare, there is no future, there will be no Europe. It seems natural for us to ask: *"If we have already got economic, monetary and labour market unity, why can't we have social convergence on this issue?"* Families of Europe hope that their cause will be listened to and promoted as a matter of priority on the European Agenda.

References

- Testa, M. R. (2006). SPECIAL EUROBAROMETER N° 253, Wave 65.1, *"Childbearing Preferences and Family Issues in Europe"*: EU25, European Commission.
- *Communication from the Commission to the European Parliament, the Council, the European Economic and Social Committee and the Committee of the Regions* (10.05.2007) - Promoting solidarity between the generations, COM (2007) 244 final.
- Communication from the Commission to the European Parliament, the Council, the European Economic and Social Committee and the Committee of the Regions (29.04.2009) - *Dealing with the impact of an ageing population in the EU* (2009 Ageing Report), COM (2009) 0180 final.
- *Draft Report on the demographic challenge and solidarity between generations* (10.05.2010) (2010/2027(INI)), European Parliament, Committee on Employment and Social Affairs.
- Cubel Sanchez, M., De Gispert Brosa, C. (2009). *"La protección de la familia en España: aún lejos de Europa"*, Documentos de Trabajo 01/09, Fundación Acción Familiar.
- Institute for Family Policies (2009). *Report on the evolution of the family in Europe*. Available from: *http://www.scribd.com/doc/22418149/Report-Evolution-on-the-Family-in-Europe-2009*.

Chapter 3: Demographic Change and the Family in Europe

Editorial

Veronika Herche
Demographic Research Institute, Hungary

We live in a rapidly changing society. Marriages and families are splitting up and new ones are forming. Forms of family life have changed and diversified over recent decades; alternative family structures and partnerships have become increasingly commonplace. Europe's youth today has a broader choice of acceptable family structures than their grandparents did. There is a trend towards family formation at later ages: studies show considerable postponement of first childbirth and first marriage since the 1970s in all European countries. Low propensities to marry are accompanied by the increasing instability of partnerships. Consequently, the number of children growing up in married-couple families has declined; single, step- or same-sex parents are no longer exceptional.

Gender roles have become less stereotyped and rigid. The rights and status of women have greatly improved during the last one hundred years. Access to education and training has increased for girls at all levels. Over the last few decades, women have gained access to managerial and other highly paid jobs long reserved for men. Parallel to the emergence of women in the labour market, men's choices, both at work and in the family, have also widened. Today's fathers play a more active and hands-on role in their children's and families' lives, although the division between breadwinner and carer roles still exists, and acts as a constraint that limits the choices of men and women.

One of the major changes of this century is that women have fewer children, and later in the life course, if at all. Having children has become a more conscious decision, amongst other reasons due to the availability of effective contraceptive methods. 'Childfreeness' is often a chosen lifestyle, for both married and cohabiting couples. Fertility rates have declined below replacement rate of 2.1 children per woman in every European country. At the same time, we are witnessing the "greying" of Europe: with rising life expectancy and declining birth rates, the age pyramid is turning into a mushroom.

According to the 2008 revision of the official United Nations population estimates and projections, the population of the 47 countries which make up Europe (according to the UN definition) is expected to decline slightly, from 732 million in 2009 to 691 million in 2050. Projections show significant changes in the distribution of the population of Europe by age group.

While the number of people aged 60 to 80 will increase by 46 per cent, the number of the oldest old will double in the next 50 years. The age group of people over 60 years will account for 44 per cent of the European population in 2050. By contrast, the number of young people aged 15 to 24 and the number of people in the main working ages (aged 25 to 59) will decrease (from 459 million in 2010 to 351 million in 2050) together with the number of children under age 15, which will decrease from 113 to 104 million. The implications of population ageing, mainly resulting from declining fertility, cannot be dismissed. According to United Nations population estimates, in the medium variant, fertility remains at 1.5 children per woman in 2045-2050, which is significantly below the replacement level of 2.1 children per woman.

Increasing longevity also contributes to population ageing. Life expectancy at birth is estimated to rise from 75.1 years in 2005-2010 to 81.5 years in 2045-2050 in Europe. Besides fertility (births) and mortality (deaths), net migration is the third driver of population change in the European Union. In 2005-2010, net migration exceeded or totally counterbalanced the excess of deaths over births in several European countries. However, the effects of immigration on population decline and demographic ageing are limited.

Demographic change poses significant challenges to society, policy makers, individuals and their families. The growing imbalance between the generations undermines the long-term financial sustainability of social systems. Is it possible to halt demographic ageing? Are incentives to increase fertility rates the answer? How far can immigration contribute to the rejuvenation of Europe's population? Then again, what do citizens want, what are people's attitudes and expectations regarding these issues?

This third chapter of *Spotlights on Contemporary Family Life* focuses on recent demographic developments and their implications for families in Europe.

In the first interview, Paul Demeny (Distinguished Scholar at the Population Council since 1989 and founder of the East-West Population Institute in Honolulu) discusses key facets of twentieth-century demographic developments in Europe, and raises some concerns about current discussions on the demographic future of Europe. One of the major topics discussed is the effectiveness of public policies enacted to solve problems arising from demographic changes.

In the second interview, Professor Herwig Birg gives his views on recent demographic trends in Europe with a focus on Germany, a country with a tradition of high welfare outlays and low fertility rates. He talks about future population and society developments in Germany, and gives some recommendations on how to manage the challenges facing policy makers and families arising from demographic changes.

The two interviews are followed by an article by Zsolt Spéder (Director at the Demographic Research Institute, Hungary). He concentrates on family changes in the new EU member states and addresses two main topics concerning family life: first, recent changes in partnership formation in the new member countries are discussed, with a focus on the relationship between non-marital and marital cohabitation; secondly, changes in the development of fertility behaviour in the region, highlighting some of the complex issues involved.

A number of recent studies have emphasised the importance of gender equality for fertility development. In the interview which follows, Livia Sz. Oláh (Associate Professor in Demography at the Stockholm University) brings us to Sweden, a country where family policies have been influenced by gender equality for decades and have made it easier for women to combine work and family life. She talks about the relationships between female labour-market attachment, policy context and fertility decisions, and underlines the complexity of mechanisms which govern the interplay between gender relations, the institutional context and individual/couple decision-making on childbearing.

The journal ends with the thoughts of Zsuzsanna Kormosné-Debreceni (Social Policy Officer at the National Association of Large Families in Hungary, Vice-President of the European network FEFAF (Fédération Européenne de Femmes Actives au Foyer)) and mother of 5 children, who emphasises the importance of having real choices and options, both for women and men. As a representative of a large family organisation, she discusses the challenges of reconciling work, family and private life in Hungary and shows how initiatives of a local NGO can support the wellbeing of families.

It is very difficult to cover the broad array of subjects and contradictory views of demographic changes in Europe in the few pages available in this chapter. But we still hope that the points of views of academic experts supplemented by the testimony of a representative of a family association will help us to understand the trends and processes behind which lie the changing faces of European families, and we very much welcome your feedback, either by e-mail or via the relevant page on the FAMILYPLATFORM website.

3.1 Demographic Changes and Challenges in Europe
Interview with Paul Demeny

Interviewed by Veronika Herche
Demographic Research Institute, Hungary

❖ *In 1922, the historian Oswald Spengler foresaw "an appalling depopulation" as one of the manifestations of the "Decline of the West". Has there been continuity in population development since the early twentieth century in Europe? Could you please give us an overview of the most important demographic shifts and trends of this region during the last century?*

To adequately describe twentieth-century demographic developments in Europe would of course take a whole book. Differences from country to country and between various social strata are just too great. Yet the key facets of the overall process can be easily summarised. Demographic change is driven by mortality, fertility, and migration. As to mortality, life expectancy at birth nearly doubled over a century: by 2000 it was slightly over 73 years for males and females combined.

The trend was steadily upward, albeit with two sharp set-backs: the first due to World War I and the influenza epidemic that closely followed it; the second, equally devastating, due to World War II: all in all, an extraordinarily positive achievement. Fertility was mainly in decline; by the 1930s some country populations and many subpopulations exhibited below-replacement levels. The post-World War II baby boom, although more moderate than in Europe's overseas offshoots - most notably in the US, was a significant but temporary reversal in the trend. In the last quarter of the century rapid fertility decline resumed and became near-universal, bringing below-replacement fertility, and often fertility deeply below replacement, in all countries of Europe by the turn of the millennium. With respect to intercontinental migration, massive European outmigration was brought to an abrupt halt by World War I. Net migratory balances in the following 40 years were very modest. But in the last decades of the century substantial immigrant flows from outside Europe have materialised, adding extra numbers mostly to the populations of the economically more dynamic countries.

Through the combined effects of these forces, as measured by any historical standard, Europe's population grew rapidly during the century: from some 390 million in 1900 to some 730 million by the year 2000. A little over half of this increase occurred in the second half of the century. The year 1900 actually provides a very arbitrary demarcation of the beginning of an epoch. That dominant trend of improving survival can be traced back to well before

1900. Fertility decline, too, had started earlier: in the case of France as far back as the second part of the eighteenth century.

For many other European countries the downward slide of fertility began in the 1880s or 1890s. The turn of the millennium, in contrast, is not a bad marker of the completion of the process of demographic transition: a transition from a combination of high mortality and high fertility to a combination of low mortality and low fertility. Europe pioneered that enormously significant historical process, setting an example for the rest of the world. The lagged response of fertility to mortality change meant that the process generated a major increase in population size. But by 2000, natural population increase - change apart from migration - came to an end for the continent as a whole. In this, too, Europe's performance prefigures what will happen - and needs to happen - elsewhere in the world. Demographic expansion cannot continue indefinitely. At some point stasis, or even modest correction through negative growth, is both inevitable and desirable. Europe is at that point now.

❖ *Europe's share of world population is in decline. Is this something to worry about?*

Normally one should not worry about things that are inevitable. Europe's loss of relative share within the world's total population has been of course steady during the past century and has been accelerating. This shift is bound to continue as far as demographers' eyes can see. In 1950, Europe's share within the global population was some 22 per cent. Today, in 2010, it may be estimated as slightly short of 11 per cent.

What will the future bring? Population projections are a risky business, but the UN's medium estimate for that share in 2050 is 7.5 per cent. That estimate assumes substantial recovery of European fertility from its current very low levels and also assumes continuing historically high net immigration – roughly 1 million persons per year. On those assumptions, Europe's 2050 population would be some 690 million (or about 40 million less than in 2010). The relative share is mostly dictated by what happens outside Europe. Europe's main concern should be how that 690 million - a very respectable number - will prosper, and how adequately it will be reproducing itself.

❖ *Europe is worried about its demographic future. Public awareness of demographic change is growing. What are the key drivers behind population ageing in today's Europe?*

The key drivers are those three factors I have just mentioned. Since population growth cannot go on forever, the convenient reference point is a population in which births and deaths roughly balance out: a stationary population or one whose underlying fertility and mortality characteristics make it headed in that direction. When just about everyone survives at least up to age 50, stationarity requires an average of very slightly more than 2 children over women's life time. When fertility falls short of that level, the base of the population pyramid narrows, thereby making the population older. And of course in modern times survival into high old age is increasingly and gratifyingly common, making an important contribution to population ageing.

❖ *The above mentioned Oswald Spengler quoted Shaw, who said the following in the section of the "The quintessence of Ibsenism" titled "The Womanly Woman" (1891):* "...unless Woman repudiates her womanliness, her duty to her husband, to her children, to society, to the law, and to everyone but herself, she cannot emancipate herself". *He continued as follows:* "The primary woman, the peasant woman, is mother. The whole vocation towards which she has yearned from childhood is included in that one word. But now emerges the Ibsen woman, the comrade, the heroine of a whole megapolitan literature from Northern drama to Parisian novel. Instead of children, she has soul-conflicts; marriage is a craft-art for the achievement of 'mutual understanding'. It is all the same whether the case against children is the American lady's who would not miss a season for anything, or the Parisienne's who fears that her lover would leave her, or an Ibsen heroine's who 'belongs to herself' - they all belong to themselves and they are all unfruitful". *What was the attitude of Europe towards population changes and their significance at the earlier 20th century?*

Such arguments, whether voiced a hundred years ago or at any time since, are little short of bizarre. Take the irrelevant contrast between the "primary woman" and the modern emancipated woman. Collective survival under conditions of high mortality of course required high fertility, an average of, say, six children per woman or even more, while today it requires two children: we are talking about completely different demographic regimes.

Shaw, a brilliant playwright, was deeply interested in social analysis and policy and wrote many penetrating pages on the subject of population. The sentence quoted by Spengler is one of those pronouncements where its author could not resist the temptation to exaggerate and to shock in the service of a good cause. No emancipation of women without repudiating womanliness and duty to children? An absurd idea. And amplifying on Shaw's false proposition, Spengler goes into an even deeper end. It is bad sociology, bad economics, and bad social psychology. Bad demography, too.

Nearly a century after his book appeared, we find that an overwhelming majority of European women - typically 80 to 90 per cent - still become mothers, and do so by choice. Do they bemoan the loss of the supposed pleasures recited by Spengler? If a large percentage of these mothers do not have a second or third child, the causes for that failure should be found in problems more real than "missing a season" and similar calamities.

❖ *In 2007, the European Commission formulated and commissioned the report on "The demographic future of Europe—from challenge to opportunity" (European Commission 2006)[1]. The paper has initiated a debate. In your article: "A clouded view of Europe's demographic future" (2007)[2], you pointed out that the "challenges and opportunities" identified in the report largely miss their target. What do you regard as the most important failings of the document?*

The report was of course a consensus document. Not surprisingly, it had a tendency to adopt a language and formulations that were calculated to smooth over differences of opinion on difficult issues or to treat major relevant subjects perfunctorily, if at all.

Does Europe need more people and if so, why? What demographic configurations justify policy interventions and what forms should they take? Why should immigration be encouraged and from what sources and in what characteristics and in what volume? Are there alternatives to garden-variety welfare state policies and what effects might such alternative approaches have on demographic behaviour? And, not least, what is the European framework in which such questions should be addressed? What are the desirable boundaries of the report's Europe – conceived at that time as the EU25 but with the prospect of enlargement? Is Europe more than a glorified customs union and is its population more than simply the sum total of the population of the member states? Or does the label "the people [in the singular] of the European Union" have a special meaning, now or in an anticipated future?

❖ *You have emphasised the obliviousness of the Commission to the issue of population size and growth. Why is it important to consider these issues when discussing Europe's demographic future?*

---

[1] See *http://ec.europa.eu/social/BlobServlet?docId=2023&langId=en.*
[2] See *http://xa.yimg.com/kq/groups/13644549/417755521/name/Paul+Demeny+on+Europe%27s+Demographic+Future.pdf.*

The Commission did touch on these issues by its comforting reference to a projected very modest decline in the size of the EU25 population: an approximate 2 per cent loss by 2050. It turns out, however, that the prospect of such near-stasis was achieved by assuming a net immigrant flow of some 40 million (plus their descendants), "conservatively estimated", as the report put it. But the size of net immigration (unlike the number of births, where grass-roots parental decisions rule, and unlike the number of deaths, where the aim of private efforts and public policies converge in the intent to keep them at a minimum) is a policy variable par excellence. How is the 40 million immigrant figure determined? It would be natural to start with population projections in the absence of immigration and address the question to what extent, if at all, the results of such projections may be problematic. Can society and the economy adjust to population decline and how? What are the disadvantages and advantages of a smaller and older population? If correction is needed, what should be the main thrust of policy intervention? These are the key questions that should have been the Commission's task to pose and answer.

❖ *One of the failings of the above-mentioned report you named is the cluelessness about fertility policy. European policy makers have not yet decided whether they should make the level of fertility rate an explicit object of government policies. What is the reason for the helplessness of governments when facing the issue of population change? Do you regard pronatalist policies as justified?*

The Commission's report carefully avoids reference to the politically incorrect term of pronatalism. It speaks, instead, of "demographic renewal", an anodyne expression signalling, it would seem, more or less the same intent. More recently, there has been some shift in terminology and explicitness. This is reflected in a hefty (almost 500 pages) United Nations report World Population Policies, 2009, that has just appeared. It characterises member state population policies and government attitudes in lapidary phrases. There is of course no EU policy on population matters; what EU members think or do is reflected in 27 country summaries, with two pages allotted to each country. Uniformity is complete with respect to "Level and concern about population age structure". For the last available year (2009) EU governments all declare that "Size of the working-age population" and "Ageing of the population" both represent "Major concerns". On "Population size and growth" and on "Fertility" there is a degree of dissonance, apparently reflecting a mixture of prevailing political and ideological positions and the most recent birth statistics. Still, not surprisingly, the majority of EU member governments view population growth as "Too low" and characterise their policy intent on population growth as "Raise". Similarly, the majority view the fertility level as "Too low" and declare that their policy on fertility is to "Raise" it. (On immigration, once again, uniformity rules: governments blandly pronounce it as "Satisfactory" and their attitude to immigration policy is to "Maintain".)

Intent and deed, however, do not easily go hand-in-hand. Policies are formulated in a political arena that seeks to weigh costs and benefits of specific measures as determined under the prevailing rules of the game. When fertility is in the neighbourhood of replacement level - neighbourhood being fairly broadly interpreted as perhaps down to a period TFR (total fertility rate) of 1.6 or 1.7 - it is difficult to argue that costly intervention (costly in terms of either public expenditure or political onus) to raise fertility is justified. Various pro-family social policies, adopted and supported for reasons other than raising the birth rate, are then presented as also being pronatalist, since they possibly have that beneficial by-product. When fertility is below replacement level by a wide margin, arguments for explicit pronatalist measures are able to command greater political support. The problem is the paucity of effective measures that have the desired effect. The main recipe is increasing socialisation of child-rearing costs and institutional arrangements that create a more child-friendly social environment and make motherhood and women's participation in the formal labour force more compatible. The record of these approaches thus far is not encouraging.

❖ *Some European countries like France, Britain or the Scandinavian countries have relatively high fertility levels, others, like most of the Eastern European countries, have lower fertility levels. What should governments of countries with very low fertility consider when contemplating what to do?*

They can certainly study apparent success stories and consider policy potentially promising approaches. But the task is not easy, since the lessons are far from obvious. It is less than clear to what extent better fertility performance in the countries mentioned are policy-related. Current fertility levels in France and in the UK, for example, are very similar, yet their social policies related to fertility are quite different. And not long ago, in the 1980s and earlier, Scandinavian countries were very much in the lower segment of European countries when ranked by level of fertility, even though their fertility-relevant social policies were regarded as the most "progressive". Recovery of fertility (or rather just some movement edging closer to replacement level) is not necessarily explainable by further reinforcement of such policies.

There are no hard-and-fast economic, social, or psychological rules governing fertility behaviour. Just a few years ago, fertility in the former East Germany was far below the level prevailing in West Germany. Today, East Germany's fertility seems to have caught up entirely with West Germany. Welcome surprises may well be in store in Eastern and Southern Europe, too, in the coming decade. Such recoveries, however natural and spontaneous, will no doubt be attributed to wise policies after the fact. Such claims will rest on weak foundations.

❖ *Many of the articles related to demographic change and policy issues contain interesting ideas but lack practical suggestions for implementing them. You suggested a couple of years ago that pension entitlements should be re-linked positively to the number of offspring produced (Demeny, 1987)? What is the main idea behind this?*

Historically, intergenerational financial exchanges and other support arrangements took place within the family. Modern industrial societies made old-age support predominantly state-organised, relying on taxing the active labour force and distributing pensions to the retired. This severs an important link between willingness to raise children and material security in old age. Re-establishing an at least partial yet significant linkage between child-rearing and entitlement to old-age support would be a potential stimulus to fertility, especially in an ageing society where government-promised pension rights come to be regarded as increasingly tenuous.

❖ *The idea of "Demeny voting" has recently been much discussed in Japan, another country with very low fertility rates. Can you explain what this voting rule means? How would its implementation affect families with children?*

In all countries, the very young - such as those under age 18 or even 20 - represent a disenfranchised population. Yet their stake in wise long-term public policies is very high (extending approximately one hundred years into the future), in contrast to the old-age population, whose relative numerical weight within the electorate is increasingly heavy, yet whose self-interested time horizon is far shorter. The young could be given electoral weight through representation by their natural or custodial parents. For example, votes for under-age girls could be exercised by their mother and for boys by their father. Other assignments of voting rights could also be contemplated. A radical version, for example, could weight all votes (including children's votes exercised by parents) by the average life expectancy at the voter's age.

Technically this (or a less discriminatory, but still age-related vote-weighting scheme) could be easily accomplished.

The constitutional and political obstacles to such a reform are of course enormous. But active advocacy of it and the ensuing debates would have a potentially strong policy-influencing effect in highlighting the inherent time-horizon bias affecting current policy decisions. I don't think of the proposal as a fertility-stimulating measure, although the recognition of the parental contribution to collective social survival would have merit and perhaps some demographic effect. The shift in the composition of representative political bodies should, however, contribute to saner policies reflecting less myopic time horizons than is common in present-day policy-making.

❖ *Last but not least, let me ask you a question concerning the FAMILYPLATFORM project. We are now at the final stage of the platform and the main goal is to develop a research agenda that encompasses fundamental research issues as well as key policy questions in order to provide an input into the EU's Socio-Economic and Humanities Research Agenda on Family Research and Family Policies. Could you name some important research needs related to demographic change, the analysis of which might help to increase the wellbeing of families across Europe?*

It would be easy to offer a long list of what ought to be researched and what policies should be contemplated. Reading the scientific output of the by now very large and very active demographic community, whether in Europe or in North America, gives a good sense of what demographers do and what policy ideas they have. Unfortunately it also gives us a sense of frustration and lack of progress: much rehashing of familiar ideas and decorating them with formal analytic virtuosity.

Instead of elaborating on this complaint, it would be wiser of me to mention just one idea which would challenge researchers in demography and also stimulate policy makers. Social policies, including attempts to deal with population issues today, originated more than a century ago from attempts to deal with issues of poverty affecting a substantial segment of the population. As advanced economies developed and incomes rose, the share of the poor within the population shrank, and material standards - nutrition, health, housing, education, spatial mobility, and leisure - improved across the board. Yet the main direction of social policies ran counter to that uplifting trend. Arrangements originally designed for the downtrodden became generalised and extended to all. Indeed, much of the redistributive function of the modern welfare state, now involving more than a third of national income, consists of taking money from the comfortably-off to reward the comfortably-off. The realistic perspective for the future, current economic set-backs notwithstanding, is further steady material improvement. Yet there is a strong likelihood that gravitation toward ever greater government-engineered redistribution of incomes will continue in the name of good causes and programmes, including programmes supposedly justified by adverse demographic developments. Does this system growing out from uplifting the downtrodden make sense in an affluent society? Is it not possible to ensure that the estimated 10 per cent of the population experiencing hard times (through no fault of their own) are properly taken care of, whilst at the same time avoiding treating the rest as if they are incapable of taking care of themselves? Demographers are well-placed to pose such unorthodox questions, since the arrangements of the modern welfare state are not exempt from the well-founded suspicion of being responsible for some untoward characteristics of contemporary society, including disorganisation of the family system and sub-replacement fertility. The prospect of fundamental social reform may seem utopian today. But the matter deserves to be considered, options need to be analysed, and radical policy alternatives need to be contemplated.

References

- Demeny, P. (1987). *Re-linking fertility behavior and economic security in old-age: A pronatalist reform,* Population and Development Review 13.1: 128-132.
- Demeny, P. & McNicoll, G. (2006). *The Political Economy of Global Population Change, 1950-2050*, a Supplement to Vol. 32 of Population and Development Review.
- Demeny, P. (2007). *A clouded view of Europe's demographic future,* Vienna Yearbook of Population Research, 2007: 27-35.
- Demeny, P. (1999). *Policy interventions in response to below-replacement fertility.* Population Bulletin of the United Nations 40/41 (special issue: *Below Replacement Fertility*): 183-193.
- European Commission. (2006). *The demographic future of Europe - From challenge to opportunity.* Communication of 12 October 2006 [COM (2006) 571].
- Shaw, G. B. (1891). *The Quintessence of Ibsenism*. London: Walter Scott.
- Spengler, O. (1922). *Der Untergang des Abendlandes. Umrisse einer Morphologie der Weltgeschichte*. München: Oskar Beck.

3.2 Are Babies Making a Comeback?
Interview with Professor Herwig Birg

Interviewed by and translated from the original German version
by Veronika Herche
Demographic Research Institute, Hungary

❖ *The widespread decline in period fertility to extremely low levels is over, claim the authors of "The end of 'lowest-low' fertility?" (published in the January 2010 issue of the quarterly "Demographische Forschung aus erster Hand" (First Hand Demographic Research)). In their analysis of recent fertility trends, Goldstein, Sobotka and Jasilioniene (2009) find a turn-around in so-called "lowest-low" fertility countries. According to their findings, Moldova was the only remaining European country with fertility rates below the 1.3 threshold in 2008, compared to 16 lowest-low fertility countries in 2002. Fears of population implosion based on continuation of fertility rates from the 1990s are no longer justified, claim the authors. Are there signs of a turnaround in population trends in Europe?*

Even before the article by Goldstein, Sobotka and Jasilioniene (2009) was released, Hans-Peter Kohler and Mikko Myrskylä of Penn's Populations Studies Center and Francesco C. Billari of the Università Bocconi published a study entitled "Advances in development reverse fertility declines" in the journal *Nature*. In the same issue of *Nature* (August 6, 2009), Shripad Tuljapurkar from the Stanford Center for Population Research published a supplementary contribution entitled "Babies make a comeback", in which he claims that in many industrialised nations including Germany, Italy and Spain, fertility levels might move back again towards the replacement level. Therefore, world population projections should be adjusted accordingly.

My critical review of the articles in *Nature* was published in the February 2010 issue of the journal *"Bevölkerungsforschung Aktuell"*. Using the example of Germany first, I revealed the non-existence of the claimed rise in fertility rates in Germany. Secondly, I showed that the arguments on which the non-existing fertility rise relies are untenable. Last but not least, I also pointed out the severe weaknesses of the methods of analysis used in the study.

There is a further example of such a glaring mistake in the history of fertility research. According to the demographic transition theory, which is, in fact, not really a theory, but rather a description of demographic tendencies, birth rates cannot sink below death rates over the long term. Since a permanent disequilibrium is excluded from this model, shrinking population can at best be temporary.

However, Germany has had a below replacement-level birth rate for almost half a century and the number of deaths has been below the number of births since 1972. The deficit of births has been increasing from year to year. In recent years, almost all industrialised countries, and increasingly more emerging and developing countries, have taken the same path: they have birth rates below the replacement level. The authors do not question these trends, and they even point out that their assumptions apply only to a few countries. If this is indeed the case, and the assumptions cannot be applied to the majority of countries, the question arises whether publishing these articles makes any sense at all.

Nothing indicates that the fertility rate will increase toward the replacement rate in the coming decades. Nor can this be forecast, since the factors outlined in my *biographic theory of fertility* (*"A biography approach to theoretical demography"*, Birg 1987, 1991) which have led to decline in fertility rates will operate in the future as well. Finally, to conclude on the articles published in *Nature*: these are entirely descriptive statistical studies with no explanatory theory behind them. This is like hiking in the woods without a compass: there is a danger of getting lost and disappearing in the undergrowth.

❖ ***There are several explanations for the decline of fertility in Europe. What do you regard as being the most important causes of the declining fertility rate?***

You will find one thing that all European countries have in common: the explosion of biographical life course alternatives since industrialisation. This is the consequence of social and cultural liberalisation on the one hand, and economic dynamism and increased welfare on the other. The biographic universe has expanded as never before in the history of mankind. In our constantly changing social, economical and cultural environment the risk of long-lasting biographic decisions increases.

Long-term biographic commitments, like choice of a long-term partner or assuming parenting responsibility are important familial virtues, but they are diametrically opposed to virtues specific to economic life, like flexibility, mobility and the constant willingness to adapt to the changing needs of the labour market. In order to avoid the risks of irreversible biographic decisions, childbirth has been postponed and is then often given up entirely. Its consequence is a decline in fertility rates; not only in Europe but all over the world. It has been faster, however, in the most dynamically growing economies. For the world population as a whole, the number of children born per woman has halved in the last fifty years.

❖ **Studies show that European couples want more children than they actually have. How can we encourage future parents to achieve their desired number of children?**

Couples do desire everything possible, not just children. Having goals is better than cultivating desires. The idea that the birth rate would increase if obstacles hindering the realisation of fertility desires could be removed is not really hitting the nail on the head.

There is a Europe-wide survey of the ideal number of children, in which people are asked what their desired number of children would be, if the state gave them all the support they wanted. On average, women and men want to have more than two children in most European countries. Nevertheless, the ideal number of children was below two in Germany and Austria. Removing obstacles and encouragement will naturally have no effect on people who do not believe that having children is self-evident and worthwhile.

❖ **Regarding birth rates, Europe is divided into two parts: a small group with comparatively high birth rates and a larger group with low fertility levels. The birth rate is exceptionally low in Germany, among the lowest in Europe. What are the main reasons for this?**

An often neglected, common denominator of the diverse reasons for the low birth rate in Germany is the historical fact that in one single century, the people of Germany have witnessed two World Wars, the hyperinflation of the 1920s, the world economic crisis of the 1930s, two dictatorships, and the 40-year-long division of the country. Furthermore, they had to cope with the social and economic transformation in the former GDR. Could you imagine that such experiences would not shake people's confidence in the future? Formation of families no longer goes without saying: it becomes a risky project.

However, it is not only negative historical experiences that have contributed to the decrease in fertility. Even positive experiences of dynamic economic growth, such as during the period of the economic boom, paradoxically had a negative impact on birth rates, mainly due to the increased parallel economic and biographical opportunity costs of having children. Therefore instead of rising, the birth rate fell to 0.7 live births per woman in Eastern Germany after reunification. I have introduced the term "*demographic-economic paradox*" for the negative relationship between economic prosperity (production) and demographic reproduction – a paradox that can only be explained by the biographical theory of fertility.

Another important historical foundation is the establishment of the universal social security system by Bismarck in the 1890s. Since then, there has been a kind of illusion of stability: people believe that they do not need to have children of their own to ensure that they will be looked after in old age or when facing severe illness. Paradoxically, only a few know that the realisation of this belief is impossible in the case of a pension system operating on the "pay-as-you-go" principle. And those who know it ignore it.

- ❖ *What can be expected in the next 50 years regarding population development in Germany? Consequently, how will the German society change?*

We are dealing with a simultaneously growing and shrinking population. The number of people aged 60 and over will increase by about 10 million to 28 million from the end of the twentieth century to 2050, and will decrease again to the initial 18 million by the end of the twenty-first century. The number of people in the 20 to 60 age group, which is the key age group for the economy, will shrink continuously up to the end of the twenty-first century; by 16 million by 2050, and then by another 10 million by the end of the twenty-first century. These estimates even take immigration effects into account.

The impact of these developments will transform Germany into a permanent social and economic construction site. The country might not be recognisable anymore. Such a development of course involves enormous political risks. The twenty-first century will be a very uncomfortable one anyway, even without any further wars. This could be unbearable for the German disposition ("Gemüt"). Nevertheless, it should be noted that immigrants with a different mentality will become the majority of the population in the younger age groups.

- ❖ *How could Germany manage the challenges of demographic change?*

Policy should pursue two strategies in parallel. First, an adaptation strategy, which would help all sectors of the economy and society to optimally adapt to the process of irreversible population ageing and decline, and to internationalisation caused by immigration. Secondly, a cause-oriented policy promoting the long-term recovery of demographic stability by achieving replacement level fertility.

Currently there is no political force in Germany that is interested in a cause-oriented strategy with long-term targets. Up until now, political parties have sold demographic problems to the voters as "opportunities of population decline" and "opportunities of population ageing". A shift has become impossible, the first party initiating a change would be voted down first. Demographic issues were not considered as important in the last

Bundestag elections. Politics seems to be bent upon continuing until reaching the end of the blind alley.

❖ *Let's talk about the sensitive issue of immigration: as the German population ages and shrinks due to a persistently low birth rate, many voices demand considering admitting more immigrants. However, there are also many concerns about immigration. Is there a way to resolve this contradiction?*

For a long time now, Germany has been replacing non-existent German children with immigrants from abroad. The annual number of immigrants has exceeded the number of births for decades. This is still so, although figures suggest that the number of emigrants temporarily exceeded the number of immigrants in 2008.

Some say that immigration problems will be defused if more people are leaving the country than are entering it. But this is not true. Even if the migration balance is zero or negative, 700,000 people or more move into the country from abroad every single year. Since they have an average length of stay of 10 years, 700,000 people need to be integrated from year to year. Simultaneous emigration does not help to solve this problem. Like in a hotel: even though the number of guests checking in and checking out is equal and the migration balance is zero, the hotel still does need a kitchen and a catering staff.

Despite the recently negative migration balance, Germany is still characterised by a policy of compensatory migration. This is, however, a dead end; the annual deficit of births will rise to 800,000 by the middle of the century. This deficit can hardly be compensated for by immigration. What really counts is the negative economic balance-sheet of immigration, not to mention the consequences for democracy if the majority of immigrants come from cultures with human rights problems.

❖ *What are the consequences and challenges of the current demographic trends for families? What challenges will arise for families as result of demographic changes?*

The Federal Constitutional Court declared German long-term care insurance unconstitutional, because in favouring childless citizens it violates the general principle of equality, the supreme principle of democracy. The court based its verdict on the demographic fact that long-term care insurance can only perform its task if people act in two areas: first, financial contributions must be paid, and secondly, children must also be born, raised and educated, in order to help preserve pay-as-you-go funding of the social insurance system with their contributions.

People who only pay a monetary contribution and still acquire the same rights to care as people who make a material, i.e. "generative" contribution, in the form of raising children, are privileged according to the court verdict. The same argumentation is applied by the court to the pension and health insurance system. The entire German social security system is therefore unconstitutional.

The result is not only a conflict of interest between contributors and beneficiaries - that means between the old and young generations, but also a conflict between the group of people with and without children within each generation. The biggest challenge for families is to bear the growing injustice resulting from privileging members of society without children, and to put up with the fact that this injustice is ignored by policy makers. In Germany those who do not have children benefit the most from this situation.

❖ *Considering current family policy practice, what are your main recommendations for Germany? How could family policy become more effective?*

In Germany, family policy focuses on people who already have children. At the same time, the proportion of people who remain childless throughout their lives within each generation is already one-quarter to one-third, and these figures are much higher among highly educated people. There should be a new policy approach for the group of childless people, so that parenthood becomes a natural element of life planning again.

Apart from that, Germany should pursue a family policy which allows both men and women to achieve their career goals and to start families simultaneously, taking France as an exemplar. So far, these two life orientations are almost mutually exclusive in Germany. Germany would probably need bipartisan coalitions to achieve the political majority needed to create and sustain effective family policies. This is still a utopian demand, but in twenty years, when demographic problems become more apparent, it might come to coalition-forming in favour of families. It might even become possible to change the constitution or the electoral law, so that parents could exercise votes for their underage children. Something similar is already included in the property rights system: children can have property rights regardless of their age, exercised by a parent until they have legal capacity.

❖ *What can and what should the economy do in order to improve living conditions for families?*

The economy is the key factor for both family formation and birth rates. It is even more important than formally responsible family policy actors such as the German Ministry of Family Affairs. In Germany, every fourth position is refilled each year. Companies should –voluntarily consider the following principle when filling vacancies, without being bound by law: where applicants have the same qualifications, priority should be given to those candidates who have family or care-giver responsibilities.

This measure would meet the important goal of creating a very dynamic economy. Since every employer has the right to define vacancy requirements in a way that only a very specific applicant profile matches them, the principle cannot be enforced legally; it could be implemented voluntarily. However, this should not be an argument against the proposal. Since all really important and valuable behaviour patterns are voluntary and are based on insight.

❖ *What societal changes being driven by demographic change should we be prepared for in tomorrow's Europe?*

Demographic development will lead to accentuation of conflict in several areas, without being driven by specific forces or groups responsible for it.

1. Since the number of elderly is growing, whereas the number of people in the middle age groups is stagnating or shrinking significantly, as in Germany, the clash of interests between social security beneficiaries and taxpayers, between the old and new generation, will intensify.
2. In addition, a conflict of interest will also emerge within each generation between individuals with and without children, because childless people are favoured by the system, as long as they only make monetary contributions but make no material, i.e. "generative" contribution to the funding of the social security system, but nevertheless acquire the same rights to care as people with children.
3. Due to the strong migration flows between communities and regions (internal migrations), regional living spaces will be split into some for winners and some for losers. The economically stronger in-migration regions will prosper at the expense of out-migration areas, both economically and demographically. Internal migration will lead to a kind of demographic colonialism, which continues the process of demographic colonialism caused by international migration, within the country.
4. At the national level, continuous immigration from abroad combined with the birth surplus of the immigrant population will result in the following change: the current majority will become the minority in the younger population groups.

References

- Birg, H. (1987). *A Biography Approach to Theoretical Demography.* IBS-Materialien 23. University of Bielefeld.
- Birg, H. (1991). *A biographic analysis of the demographic characteristics of the life histories of men and women in regional labour market cohorts as clusters of birth cohorts*, in Becker, H. A. (ed.) *Life histories and generations, Vol. 1.*, University of Utrecht, 145-182.
- Birg, H. (1995). *World Population Projections for the 21st Century - Theoretical Interpretations and Quantitative Simulations.* Frankfurt: Campus Verlag and New York: St. Martins Press.
- Birg, H. (2004). *Die Weltbevölkerung - Dynamik und Gefahren.* München: C. H. Beck.
- Birg, H. (2005). *Die demographische Zeitenwende - Der Bevölkerungsrückgang in Deutschland und Europa.* München: C.H. Beck.
- Birg, H. (2006). *Die ausgefallene Generation - Was die Demographie über unsere Zukunft sagt.* München: C.H. Beck.
- Birg, H. (2010). Review of Mikko Myrskylä, Hans-Peter Kohler and Francesco C. Billari (2009). *Advances in development reverse fertility declines.* In Nature 460.6: 741-743 and Tuljapurkar, S. (2009). *Babies make a comeback.* In Nature 460.6: 693-694. Bevölkerungsforschung Aktuell, 7.2. Wiesbaden: Bundesinstitut für Bevölkerungsforschung. Retrieved from: http://www.bib-demografie.de/nn_750528/SharedDocs/Publikationen/DE/Download/Bevoelkerungsforschung__Aktuell/bev__aktuell__0210,templateId=raw,property=publicationFile.pdf/bev_aktuell_0210.pdf.
- Goldstein, J.R., Sobotka, T. & Jasilioniene, A. (2009). *The end of "lowest-low" fertility?* Population and Development Review 35.4: 663-699.
- Goldstein, J.R., Sobotka, T. & Jasilioniene, A. (2010). *Geburtenraten in vielen Industriestaaten steigen wieder: Eine nachhaltige Trendumkehr scheint möglich*, Bevölkerungsforschung Aktuell 7.1. Wiesbaden: Bundesinstitut für Bevölkerungsforschung. Retrieved from http://www.demografische-forschung.org/archiv/defo1001.pdf.
- Kohler, H-P., Myrskylä, M. & Billari, F. C. (2009). *Advances in development reverse fertility declines.* Nature 460.6: 741-743.
- Tuljapurkar, S., *Babies make a comeback.* Nature 460.6: 693-696.

3.3 Family Changes in the New EU Member States[1]

Zsolt Spéder
Demographic Research Institute, Hungary

Two decades ago, profound changes took hold in the former socialist countries. In many respects, these changes can be regarded as having come to a close: regular free elections regulate the changes of governments, a characteristic feature of liberal democracies; the (free) market integrates national economies; and the countries of Central and Eastern Europe have joined the European Union. Regarding the family, several specific changes have also taken place. However, they did not commence at the same times, nor did they proceed at the same speeds. These changes in the family have so far not come to a close. We cannot confirm with any certainty that identical models will emerge in all of the former socialist countries. But based on empirical findings, in this contribution we highlight and compare selected aspects of family changes in the new member states of the European Union.

1. Introduction: questions and context

Several reasons underlie our intention to focus on family changes in the former Socialist countries among the new EU member states. On the one hand, everyday life in these countries was ruled by very similar, though not identical social forces. Since 1945 (earlier in the Baltic region), the populations of these states lived in a redistributive system that profoundly affected and constantly restructured everyday life. From 1989/90 onwards, these countries oriented themselves on the Western European model - taking over the entire political system, developing the conditions of a market economy and privatising state property - developments that culminated after the turn of the millennium in the accession of ten (8+2) countries to the European Union (Adamski *et al.*, 2001). The transition to a market economy happened at a time when globalisation processes intensified in the West. Ultimately, the elimination of borders gave the green light for cultural integration. The structural circumstances in which people live and need to find their bearings appear at first glance to resemble each other and to show parallel developments in the long term. Perhaps we might assume that these countries have also undergone the very same changes with regard to the family?

Indeed, there are arguments that suggest distinct differences in family changes between accession countries. First, the Baltic countries were

---

[1] This is a shortened, modified and updated version of the paper by Spéder (2009).

integrated in the Soviet system more profoundly and over a longer period. The Church's role in the new member states clearly varies to this day, and its influence on family life still makes itself felt; these and other processes may give rise to dissimilar family tendencies. Secondly, according to some analyses (see King/Szelényi, 2005) the former communist countries chose different paths for the (re-)introduction of capitalism, thus giving rise to different structures. Lastly, we assume that long-term cultural developments, as well as patterns and developments of the welfare state, affect family changes. John Hajnal (1965) identified the continuing differences between West and East by analysing long-term development tendencies for marital age. Reher (1998) emphasised North-South differences and Mayer (2001) pointed out that different welfare state systems have a crucial impact on familial circumstances. These arguments justify a comparison of tendencies in family changes in those former Socialist countries that acceded to the European Union a short while ago. They are simultaneously similar and dissimilar, and allow us to identify specific features.

The present paper is devoted to two main topics, with a closer look at a number of selected aspects of family life based on survey findings. These topics are as follows:

1. The interrelation between non-marital cohabitation (hereafter referred to as "cohabitation") and marital cohabitation (hereafter referred to as "marriage").
2. Development of fertility, i.e. fertility rates, attitudes and social norms towards childbearing.

Our analysis is based on several data sources and various types of data. The basic tendencies are shown by means of vital statistics. Two available sets of data - the data of the first wave of the "Generations and Gender Survey" (cf. Vikat *et al.*, 2007), and the "Timing of the Life" module of the third wave of the "European Social Survey" (ESS) - allow us to conduct further analyses.

2. Forms of partnership: marriage and/or cohabitation

2.1 General tendencies

The past two decades have brought profound changes in partnership formation in the NMSs (New Member States): the popularity of marriage has diminished considerably; cohabitation has become a universal phenomenon; the locating of first partnership and marriage in the life-course has changed; the stability of partnerships has not remained unaffected either (see Bukodi,

2004; Spéder, 2005; Hoem *et al.*, 2007; Sobotka/Toulemon, 2008). These changes resemble those that have taken place in the Western countries since the middle of the 1960s (Lesthaeghe, 1996; Kiernan, 2000a). However, we must ask ourselves whether such changes occurred in the same way in all of the countries mentioned. Another issue is the nature of the competition between marriage and cohabitation in the new member states.

The traditional indicator - total first marriage rates (TFMR) - reveals that the willingness to marry has declined considerably and that first marriage, even in the NMS, clearly takes place at a later time than before (cf. Billari, 2005; Lesthaeghe/Moors, 2000). However, differences in total first marriage rates are found in the extent of changes and the initial and final levels. Rates decreased from 0.75 to 0.44 in Hungary, from 0.82 to 0.69 in Romania, and from 0.87 to 0.58 in Poland.

The motives and causes of postponement of union formation concur with those commonly found in the Western countries. These are as follows: expansion of higher education, later entry into working life, growing economic insecurity, changes of norms and values, increasing prevalence of cohabitation.

Essential differences between Western and Eastern countries are also found in the explanations given for postponing first marriage ("subjective causes"). Whereas in Western countries people tend to name values and subjective orientations, such as striving for personal freedom, a declining appreciation of marriage and higher social acceptance of cohabitation, it is material factors – the housing shortage, low incomes, labour market problems - that play a crucial role in postponing marriage in the former Socialist countries (Pongrácz/Spéder, 2008).

2.2 Popularity of cohabitation as partnership form

It is not without reason that several researchers claim that cohabitations - even though in different ways and to different extents - have become an inevitable form of partnership and/or integral part of partnership careers (Toulemon, 1997; Kiernan, 2002b; Vaskovics, *et al.*, 1997). What can we say about the actual popularity of cohabitation in the countries surveyed?

Since the statistics do not provide sufficient information, we need to extend our scrutiny to survey findings. According to the data of the third wave of the European Social Survey (ESS), fairly substantial proportions of young adults (aged 21-35) living in steady partnerships prefer cohabitation. Variations between NMSs are significant: while in Poland and Slovakia very few young couples cohabit, proportions are fairly large in Estonia and Slovenia. The wide variance in the prevalence of cohabitation is typical of both 'old' and 'new' member states. In contrast, the proportions of people cohabiting are small in the Ukraine and Russia. Thus, it is not surprising that acceptance of cohabitation is very high, with only few people disapproving. This coincides with the findings of previous surveys (see Spéder, 2006).

Table 1: Partnership forms of young adults with a cohabiting partner (aged 21-35), 2006

	% living in a non-marital partnership	in a marriage
Western countries		
Norway	57.1	43.9
France	50.0	50.0
Germany	37.3	62.7
Spain	33.2	66.8
NMS countries		
Estonia	47.9	52.1
Poland	10.8	89.2
Slovakia	19.8	80.2
Hungary	33.7	66.3
Slovenia	56.3	43.7
Bulgaria	27.2	72.8
Eastern countries		
Russia	12.7	87.3
Ukraine	2.3	97.7

Source: *own calculations based on the "European Social Survey", 3rd wave (2006).*

However, partnership practice and evaluation of partnership forms by the public do not coincide fully. Whereas half of the young adults living in a partnership in Estonia but a mere eighth in Russia are in an cohabitation, these two countries show hardly any differences in the preferences towards non-marital cohabitation.

2.3 Dissolution of partnerships: growing risk

Dissolution of marriage - the high and still growing readiness to divorce - is the most important source of partnership instability. While it is true that the readiness to divorce has increased in general, total divorce rates (TDR) barely changed between 1988 and 2003 in three NMS countries (Estonia, Lithuania and Romania). In Lithuania, TDR even fell by 0.13. The extent of increases also differs among countries. Hungary is on top with 0.17, followed by the Czech Republic. Accordingly, the within-country variance of TDR, the dissolution risk of a marriage is greater in 2003 than in 1988: the 2003 rates vary between 0.2 and 0.5. TDR figures reveal that marriages in the Czech Republic have a likelihood of ending in divorce of 0.48, whereas in Poland the corresponding probability is one-fifth (0.20). The findings show distinct differences between the new member states in certain areas of family life.

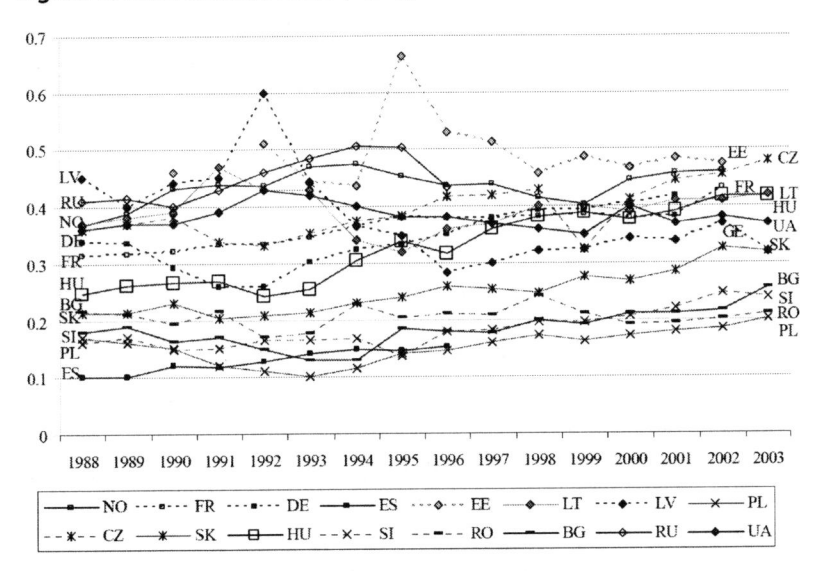

Figure 1: Total divorce rates (TDR), 1988-2003

Source: *Recent demographic developments in Europe (2004)*.

The diffusion of cohabitation does not decrease the instability in partnerships either. Since persons living in cohabitation tend to dissolve their partnership more easily than married people, and since cohabitation is spreading inexorably, total rates of dissolved partnerships are increasing.

In summary: profound changes have occurred in the new member countries of the European Union, and they are still under way. Young adults choose the binding and institutionalised form of marriage less frequently and/or at a later time, preferring cohabitation as a pre-marital or alternative form of partnership. There can be no doubt that both marriage and cohabitation are subject to a change of relevance – and most respondents have come to accept this fact. Marriage is a desired life goal for most people, but instead of being an exclusive institution, it has become an idealised form of partnership "re-charged" with specific values.

3. Tendencies in childbearing: facts and arguments

It is not possible to give a short yet precise description and final summary of changes in the development of fertility behaviours[2]. Researchers studying fertility behaviour are confronted with an equally difficult but just as exciting challenge as the sociologists who investigated social transformation processes at the beginning of the 1990s. They would have had to understand and explain changing processes at a time when they were still in progress. We have no choice but to highlight some aspects of fertility behaviour and to try to understand them. The fundamental tendencies will be described below with the aid of the total fertility rates (TFR). Subsequently, we will deal with the postponement of the (first) birth to a later age. We will specify the underlying causes, describing and discussing some characteristic attitudes and norms. We will then conclude with the recurring problem of reconciling job and family.

3.1 Fundamental tendencies and characteristics

Total fertility rates (TFR) reveal that the individual new member states have followed similar paths. We find a rapid decline in fertility in all new member states. According to Sobotka (2008), the extent and rate of decline were greater than in the West European countries, where these changes had occurred at an earlier time. If we were to characterise the changes by just this one indicator - incidentally one of the most important indicators of fertility - we would find that the former Socialist countries now yield a somewhat more uniform picture than they did in the late Socialist period before transformation.

Changes in the average age of mothers at first birth (MAFB) also reveal an identical trend: women are increasingly giving birth to their first child at a later age. And there is no sign as yet that this trend will come to a standstill.

This is the most useful indicator for us to understand the nature of the change in fertility behaviour, as it shows postponement of the first birth. It is not an unfamiliar process, having first appeared in Western societies a long time ago. Yet we do not find it in Russia and Ukraine, which means that it is not necessarily linked to the transformation process, i.e. the transition from Socialism to a market economy (cf. Adveev, 2003; Perelli-Harris, 2005).

---

[2] A concise, detailed description and analysis of the trends was recently carried out by Sobotka (2008).

Figure 2: Total fertility rates (TFR), 1988-2009

Source: *Eurostat.*

Figure 3: Average maternal age at first birth (MAFB), 1988-2003

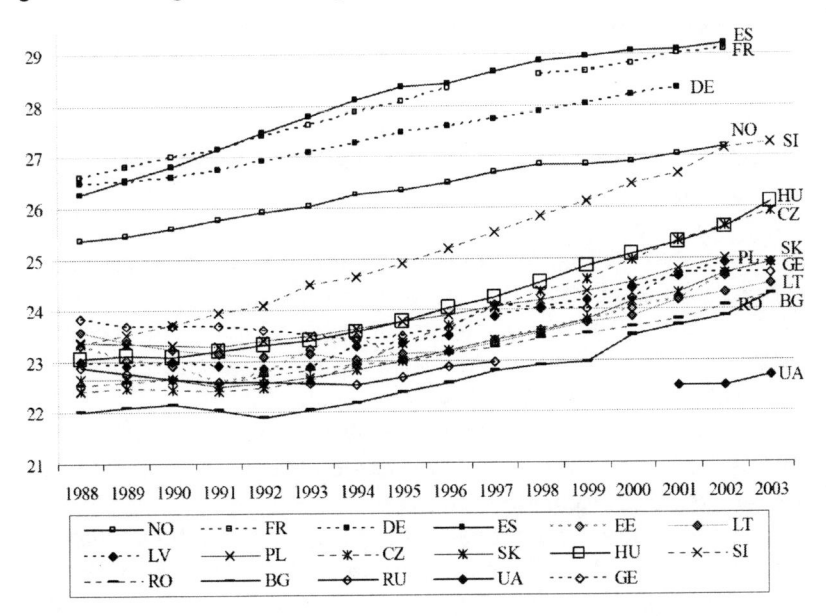

Source: *Recent demographic developments in Europe (2004).*

There is no doubt about some of the causes of postponement, first of all the expansion of schooling: the number of university students among 20 year-olds is currently three times the number before political transformation. Labour market integration of young adults comes second, which implies a delayed transition from schooling to the workplace. As shown by Mills and Blossfeld in their comprehensive studies (2005), globalisation has also played a key role, and we need to underscore that globalisation had it easier in the new member states than in the Western countries: shaping the weak market economies to suit globalisation concepts was fairly simple, but led to greater insecurity compared to Western countries. It is well known that this insecurity at transition from education and training to workplace delays all other events across an individual's lifetime, including the decision to have a child. Finally the increasing prevalence of cohabitation and its characteristic features, i.e. its transitional character, the greater readiness of the partners to dissolve it, and the lower propensity to fertility, contribute to the postponement of first births, a delayed transition to adulthood, as well as other postponement tendencies.

Although the two indicators mentioned above yield important information about trends in fertility behaviour, they provide only limited insight into this behaviour. They do not tell us how many children a woman or family will have nor anything about the prevalence of childlessness, how often and how long children will live with a single parent, nor anything about the relevance of children in the lives of parents. In the following section, we will refer to the survey data to find answers to some of these questions.

3.2 Opinions, anticipations, deliberations and plans with regard to children

As a first question, let us ask whether children are still major life goals. In the Generations and Gender Survey, we asked whether having a child is essential for a fulfilled life. We proceeded on the assumption that the replies would yield information on the significance of motherhood and fatherhood as human life goals. We expected gender-specific distributions, i.e. a higher relevance of children for female than for male identity, but found only slight differences between countries (with the exception of Germany, where childlessness is a relatively long tradition).

What are the differences between countries in view of the ultimate family size and intentional childlessness? A recovery in the birth rate is possible only in the case of sufficiently high ultimate family size. According to Goldstein and his associates, the German-speaking European countries did not only experience a "sub-replacement level of fertility" but an alignment of the ultimate fertility goals to this level (Goldstein/Lutz/Testa, 2003). And these figures are far removed from the "magical two". So we are left with the question of whether expectations in the new member countries are also oriented to actual fertility rates.

Eurobarometer data suggest that this is not yet the case. In eight out of ten countries, fertility fluctuates between 1.9 and 2.1. Only two countries, the Czech Republic and Romania, have rates below 1.9. Aside from this, it is a remarkable phenomenon that ultimate fertility goals in France and Norway and Spain far exceed the 2.1 limit.

Table 2: Ultimate family size of young adults (aged 18-34) in various European countries, 2001

Countries	Ultimate family size
NMS countries	
Bulgaria	2.00
Estonia	2.03
Lithuania	2.09
Latvia	2.06
Poland	2.14
Czech Republic	1.82
Slovakia	1.97
Hungary	2.05
Slovenia	1.96
Romania	1.70
Western countries	
Finland	2.41
France	2.48
Germany	1.64
Austria	1.73
Spain	1.97

Source: *own calculations, Eurobarometer 59.2, Candidate Countries Eurobarometer (2002).*

Researchers can see childlessness as a special type of fertility goal. Put very simply, childlessness is the consequence of conscious thought and decisions, a side-effect of postponing behaviour. Alternatively, it can also result from the early dissolution of a partnership, etc. Our question is how people rate childlessness, and whether it comes with open or covert sanctions, such as disparaging remarks, gossip or non-verbal reactions. In order to record and describe these sanctions, we will use data compiled by the European Social Survey (ESS).

The populations of the Western countries definitely have no negative views of childlessness. Since intentional childlessness is not widespread (except in Germany), this attitude might be a sign of tolerance rather than acceptance. In the NMS countries, the proportion of disapproval is much higher, but with considerable differences between these countries. While in Slovenia 40 per cent of respondents see childlessness as a negative phenomenon, the rate is 80 per cent in Bulgaria. In the Western countries, more people expect negative sanctions than disapprove of childlessness themselves.

Consideration of anticipated benefits and drawbacks of having a child is an integral part of all theoretical deliberations that lend a kind of rationality to fertility behaviour[3]. The anticipated positive and negative consequences of a (potential) child are measured by means of the following set of questions: "If you were to have a/another child during the next three years, would it be better or worse…:

- For your employment opportunities?
- For your spouse's employment opportunities ?
- For your financial situation?
- For the joy and satisfaction that you get from life?
- For the possibility to do what you want?"

The anticipation of positive consequences of a potential birth contributes to the emergence of the wish to have children, while anticipated negative consequences may prevent people from having a child. Anticipations are homogenously negative in regard to women's employment opportunities, with hardly any country-specific differences: in all countries, people clearly and exclusively anticipate negative consequences. A more positive dimension is the expectation of positive changes in joy in the case of a birth. France and Georgia have the highest number of persons envisaging that the birth of a child will increase their joy and satisfaction, while Germany has the lowest. In the NMS countries, anticipation of positive consequences is greatest in Hungary. Of course, there are distinct differences between childless persons and families with children: childless persons tend to anticipate positive consequences, families with children tend to anticipate more negative consequences.

---

[3] The GGS data set mentioned here is based on the theory of planned behaviour as developed by Ajzen (1988). On the advantages of Ajzen's theory see Vikat et al. (2007).

Figures 4a and 4b:

Women's anticipations:
Anticipated negative consequences in the labour market in the case of a birth in the next three years

Women's and men's anticipations:
Anticipated positive changes in joy in the case of a birth in the next three years

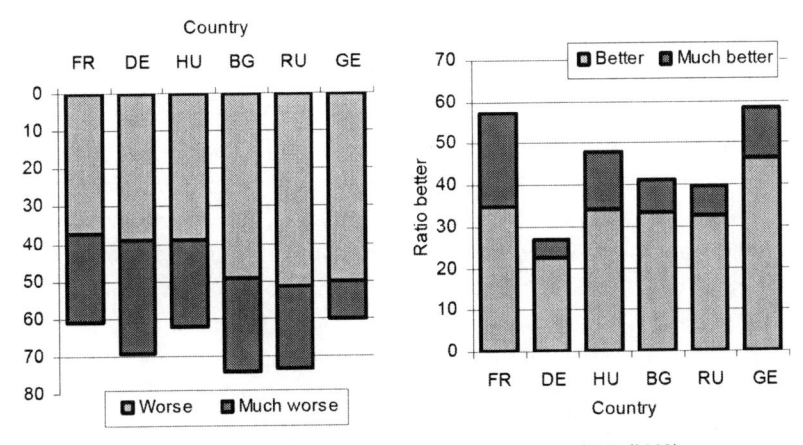

Source: *own calculations, "Generation and Gender Survey", 1st wave (2001/2002).*

In our analyses, we also tried to identify the impact of labour market status (Spéder/Kapitány, 2007). We wanted to reveal the influence of labour market status on the emergence of the wish to have children as well as its realisation. These analyses indicate that people with medium qualifications who can achieve only weak status in the labour market have the lowest chances of fulfilling their fertility goals. In Hungary, to remain childless forever is not a fate that threatens higher income groups, although having children comes at a high cost in terms of opportunities. By contrast, the extent of fulfilling fertility goals is lowest in the middle positions. Perhaps they are the ones who are most at risk of losing a great deal. Above-average earners can use their accumulated available capital to make up for sudden financial losses. For the lower classes in Hungary, social benefits can only complement or partly compensate for income losses. These findings direct our attention to the problems of reconciling job and family.

3.3. Weak reconciliation of work and family in the post-communist countries

How can it be that reconciliation of employment and family becomes a problem in the former Socialist countries, where universal gainful employment of women characterised the labour market for half a century? Mostly because labour market conditions have undergone fundamental changes (cf. Spéder/Kamarás, 2008). We can give only a brief summary of these fundamental changes:

1. These countries were caught by globalisation in a situation of insufficiently developed domestic markets (cf. Mills/Blossfeld, 2005). The working population, including women, were unable to defend their rights adequately. Having played merely a formal role during Socialism, trade unions were unable to position themselves afresh, and so they lacked clout. The state, too, proved too weak to promote the interests of employees.
2. Part-time employment rates in these countries are amongst the lowest in Europe. If both women and men work full-time (eight or more hours a day), fulfilling their familial tasks is difficult.
3. Although employers must re-employ their employees after maternity leave, practical implementation of this right is inadequate.
4. According to our studies, women with middle qualifications and weak labour market positions are mostly affected by this problem, while university graduates and highly qualified experts are in a strong position in the labour market thanks to their skills and knowledge. At the other end of the spectrum, persons with low qualifications have easier access to marginal positions in the labour market and receive the largest share of government benefits to compensate for most of their income losses.

All items mentioned above illustrate the persisting problem of reconciling job and family.

3.4 Births out of wedlock

The varying distributions of cohabitation reveal distinct differences between the new member states in regard to births out of wedlock; in addition, evaluations of this situation in these countries are not homogenous. The largest differences regarding family tendencies can be found in this case. Not only is Estonia in the vanguard, but its rate - close to 60 per cent (58.2 per cent) in 2006 - is also noteworthy compared to the rest of Europe. High rates (over 40 per cent) are also found in Latvia, Slovenia and Bulgaria, three countries that differ in many respects. Poland is at the other end of the scale (together with Ukraine), with figures of around 20 per cent. These countries, seemingly homogenous from the viewpoint of total fertility rates, showed large differences in regard to partnerships.

Of course not all non-marital children are born into a cohabiting situation and subsequently raised by two parents. The proportions of children born in single-parent families differ widely in Western Europe. In the United Kingdom this proportion is quite high. According to the censuses around the millennium, the proportion of children below age one living in a single-parent household varies between 5 and 20 per cent in the NMS countries.

The rating of out-of-wedlock births by the respondents corresponds to the former's prevalence, thus showing a distinct variance. Nevertheless, there are a few discrepancies: acceptance of cohabitation is highest in Hungary rather than in Estonia. Although cohabitation prevalence in Hungary is only in the medium range, it is tolerated by almost all respondents.

It would be important to know more about the socio-structural embedding of children born out of wedlock. All we know is that Hungary's model is similar to that of the United Kingdom but differs from the model of Scandinavian countries. In these two countries, births out of wedlock tend to be a characteristic feature of specific social groups, i.e. the lower ranks of society. While cohabitations are found in all ranks of society, having a child within this type of partnership is more characteristic of the lower ranks.

If we summarise fertility tendencies, the NMS countries are quite similar. Postponement of first births is characteristic of them all, which implies total fertility rates around 1.3. This is indicative of the fact that societal transformation - institutional/structural changes - might play a pivotal role in fertility processes. These countries started from a nearly identical institutional/structural position (the late Socialist period), set themselves identical goals (catching up with Western European societies) and tried to find their bearings in the same circumstances (increasing globalisation). All of this helps explain the similarities. However, it would be premature to assume that people in these countries will behave identically in the future with regard to these issues (cf. Coleman, 2004). We must not forget that:

a. we are currently in the phase of the (second) demographic transition, i.e. a phase that has not yet come to a close;
b. low total fertility rates are the result of transitional behaviour, because people do not want children now: they cannot be seen as an indicator of a new and stable fertility behaviour;

c. although all these countries became integrated in the globalised market economy, their family-related institutions show considerable differences: so far, we cannot say with any certainty towards which Western welfare system they are moving;
d. we must not underestimate the impact of cultural values and traditions that persist despite growing European unification.

4. Summary

Our short summary of family changes in the new member countries shows that they are characterised by both homogeneity and lack of homogeneity. Whilst similarities are more likely to be found in fertility behaviour, there appear to be more differences regarding partnership behaviour.

Whereas Western countries have taken quite divergent paths, with currently contrasting fertility behaviour (e.g. the differences between Scandinavian and Southern countries), the NMS countries surveyed appear to follow close if not identical transformative paths (Frejka/Sobotka, 2008). However, we cannot rule out the possibility that this pronounced uniformity is only due to the fact that these countries have not yet decided which new patterns to choose, and that concurrence is found only in the "decision not to have children". In Western Europe, the relationships between fertility, marriage and partnership behaviour have changed: unlike previously, the interrelation between total fertility rates and total first marriage rates has turned negative (Billari, 2005; Sobotka/Toulemon, 2008). By contrast, the countries surveyed by us also have low marriage popularity levels, but the proportion of children born out of wedlock varies from country to country. The differences result from different partnership behaviour (form of first partnership, duration of partnership, divorce risk).

It is obvious that these issues require additional research. The NMSs are not drifting unequivocally towards a Western European family pattern. Hence our assumptions are closer to those researchers who suggest there are several different patterns, rather than a single uniform European pattern (Billari/Wilson, 2001; Mayer, 2001; Reher, 1998). We cannot rule out the possibility that these countries, are "designing" a specific family pattern which is new (in European terms). However, there are three statements that we can make with certainty. First, the change of political regime was followed by profound changes in familial circumstances. While social change is an integral part of modern societies, as invariably stressed by Zapf in his works (Zapf, 1996), these familial changes have been faster and more intensive than they would have been under normal conditions. Secondly, we can claim with certainty that this change has not come to a close, at least so far.

Thirdly, we can be sure that in spite of having adopted many West European institutions, the NMSs have increased European family diversity.

References

- Adamski, W., Machonin, P., Zapf, W. & Delhey, J. (eds.) (2003). *Structural change and modernization in post-socialist societies*. Hamburg: Krämer-Verlag.
- Avdeev, A. (2003). *On the way to one-child family. Are we beyond the point of return? Some considerations concerning the fertility decrease in Russia*. In Kotovska, I.E. & Józwiak, J.(eds.) *Population of Central and Eastern Europe. Challenges and opportunities*. European Population Conference. Warsaw: Statistical Publishing Establishment.
- Ajzen, I. (1988). *Attitudes, Personality and Behaviour*. United Kingdom: Open University Press.
- Billari, F. C. (2005). *Partnership, childbearing and parenting: Trends of the 1990s*. In Macura, M., MacDonald, A.L. & Haugh, W. In *The New Demographic Regime. Population Challenges and Policy Responses*. New York and Geneva: United Nations: 63-94.
- Billari, F. C. & Wilson, C. (2001). *Convergence towards diversity? Cohort dynamics in the transition to adulthood in contemporary Western Europe*. Rostock: Max-Planck Institute for Demographic Research (Working paper 2001-039).
- Bukodi, E. (2004). *Ki, kivel (nem) házasodik? [With whom are you married (or not)?]*. Budapest: Századvég.
- Coleman, D. (2004). *Why we don't have to believe without doubting in the "Second Demographic Transition" – Some agnostic comments*. Vienna Yearbook of Population Research 2004: 313-332.
- Frejka T. & Sobotka, T. (2008). *Overview Chapter 1: Fertility in Europe: Diverse, delayed and below replacement*. Demographic Research 19: 15-46. http://www.demographic-research.org/Volumes/Vol19/3/default.htm.
- Goldstein, J., Lutz, W. & Testa, M. (2003). *The emergence of sub-replacement family ideals in Europe*. European Demographic Research Papers, VID Vienna, 27.
- Hajnal, J. (1965). *European marriage patterns in perspective*. In Glass, D.V. & Eversley, E.C. (eds.) *Population in history*. Chicago: Edward Arnold, 101-143.
- Heuveline, P., Timberlake, J. M. & Furstenberg, F. F. Jr. (2003). *Shifting childrearing to single mothers: Results from 17 Western countries*. Population and Development Review 29.1: 47-71.
- Hoem, J. A., Jasilioniene, A., Kostova, D. & Muresan, C. (2007). *Traces of the*

Second Demographic Transition in selected countries in Central and Eastern Europe: Union formation as demographic manifestation. Rostock: Max Planck Institute for Demographic Research (Working paper 2007-026).
- Kiernan, K. (2000a). *European perspectives on union formation.* In Waite, L. (ed.) *The ties that bind.* New York: de Gruyter, 40-58.
- Kiernan, K. (2002b). *Unmarried cohabitation and parenthood: here to stay? European perspectives.* Paper presented at the *Conference on Public Policy and the Future of the Family.* October 25, 2002.
- King, L. B. & Szelényi, I. (2005 [1994]). *Post-communist economic systems.* In Smelser, N.J. & Swedberg, R. (eds.) *The handbook of economic sociology.*, 2nd ed. Princeton: Princeton University Press, 205-229.
- Lesthaeghe, R. (1996). *The second demographic transition in Western countries: An interpretation.* In K. Oppenheim Mason & A.-M. Jensen (eds.). *Gender and family. Change in industrialised countries.* Oxford: Clarendon Press, 17-62.
- Lesthaeghe, R. & Moors, G. (2000). *Recent trends in fertility and household formation in the industrialized world.* Review of Population and Social Policy 9: 121-170.
- Mayer, K. U. (2001). *The paradox of global social change and national path dependencies: Life course patterns in advanced societies.* In Woodward, A.E. & Kohli, M. (eds.) *Inclusions, exclusions.* London: Routledge, 89-110.
- Mills, M. & Blossfeld, H. P. (2005). *Globalization, uncertainty and the early life course: A theoretical framework.* In Blossfeld, H. P. et al. (eds.) *Globalization, uncertainty and youth in society.* London: Routledge, 1-24.
- Perelli-Harris, B. (2005). *The path to lowest-low fertility in Ukraine.* Population Studies 59.1: 55-70.
- Pongrácz, M. & Spéder, Z. (2008). *Attitudes towards forms of partnership.* In Höhn, C., Avramov, D. & Kotowska, I. E. (eds.) *People, population change and policies.* Dordrecht: Springer Verlag, 98-112.
- Reher, D. S. (1998). *Family ties in Western Europe: Persistent contrasts,* Population and Development Review 24: 203-234.
- Sobotka, T. (2008). Overview Chapter 6: *The diverse faces of the second demographic transition in Europe.* Demographic Research 19: 171-224. http://www.demographic-research.org/Volumes/Vol19/8/default.htm.
- Sobotka, T. & Toulemon, L. (2008). Overview Chapter 4: *Changing family and partnership behaviour: Common trends and persistent diversity across Europe.* Demographic Research 19: 85-138. http://www.demographic-research.org/Volumes/Vol19/6/default.htm.
- Spéder, Z. (2005). *The emergence of cohabitation as first union and some neglected factors of recent demographic developments in Hungary.* Demográfia Special English Edition. Budapest: DRI.

- Spéder, Z. (2006). *Childbearing Behavior in the New EU Member States: Basic Trends and Selected Attitudes.* In Lutz, W., Richer, R. (eds.) *The New Generations of Europeans. Demography and Families in the Enlarged European Union.* London: Earthscan, 59-82.
- Spéder, Z. & Kapitány, B. (2007). *Gyermekek: vágyak és témnyek.* Budapest: KSH Népességtudományi Kutatóintézet (Életünk fordulópontjai Műhelytanulmányok 6. szám), 186.
- Spéder, Z. & Kamarás, F. (2008). *Hungary: Secular fertility decline with distinct period fluctuations.* Demographic Research 19: 599-664.
- Spéder, Z. (2009) *Familiale Entwicklungsverläufe in den neuene EU-Mitgliedstaaten,* in Kapella, Rille-Pfeifer, Rupp, Schneider (Hrsg.). 2009. *Die Vielfalt der Families.* Opladen: Verlag Barbara Budrich: 391-420.
- Toulemon, L. (1997). *Cohabitation is here to stay.* Population: An English Selection 9: 11-46.
- Vaskovics, L. A., Rupp, M. & Hofmann, B. (1997). *Lebensverläufe in der Moderne:* Nichteheliche Lebensgemeinschaften. Eine soziologische Längssschnittstudie. Opladen: Leske + Budrich.
- Vikat, A., Spéder, Z., Beets, G., Billari, F.C., Bühler, C, Désesquelles, A., Fokkema, T., Hoem, J.M., MacDonald, A., Neyer, G., Pailhé, A. Pinnelli, A. & Solaz, A. (2007). *Generations and Gender Survey (GGS): Towards a Better Understanding of Relationships and Processes in the Life Course.* Demographic Research 17: 389–440. http://www.demographic-research.org/Volumes/Vol17/14/default.htm.
- Zapf, W. (1996). *Die Modernisierungstheorie und die unterschiedlichen Pfade der gesellschaftlichen Entwicklung.* Leviathan 24: 63-77.

3.4 Childbearing in a Gender-Equal Society
Interview with Livia Sz. Oláh

Interviewed by Veronika Herche
Demographic Research Institute, Hungary

❖ **Europe has experienced a strong increase in female labour force participation in the past decades, even in the main childrearing ages. How far are changes in female labour force participation related to recent fertility developments in Europe?**

Women's engagement in paid work has increased substantially across Europe, as seen in Figures 1 - 4. Currently, about 80 per cent of women in the main childrearing ages participate in the labour market in Northern and Western European countries, around 70 per cent in Central-Eastern Europe and over 60 per cent in Southern Europe. In family research it has been argued that such high levels of female employment have become possible due to small family sizes, in that having relatively few children makes it possible for women to engage in other activities beside childrearing and housework, such as paid work.

As the ideal of individualism has spread across modern societies, aspirations of self-realisation beyond the home sphere have become more and more common among women as well. At the same time, the development of contraceptive technology has provided couples and women as individuals with highly efficient means of birth control (especially the pill). A contested question is then whether women limit the number of children they choose to have in order to participate in the labour force, and whether and/or to what extent the policy context matters for such decisions.

Figures 1-4: Female labour force participation rates at ages 25-49 years in selected European countries, 1983-2008

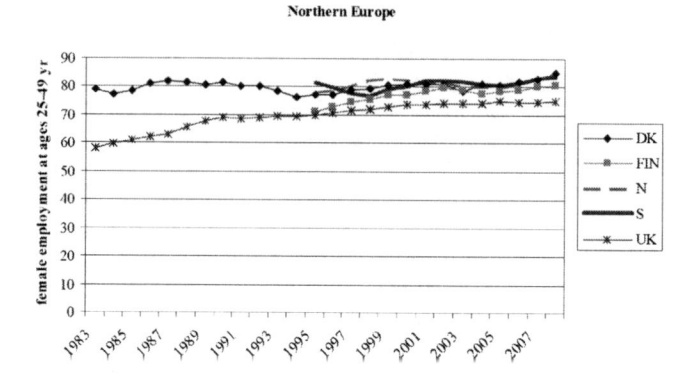

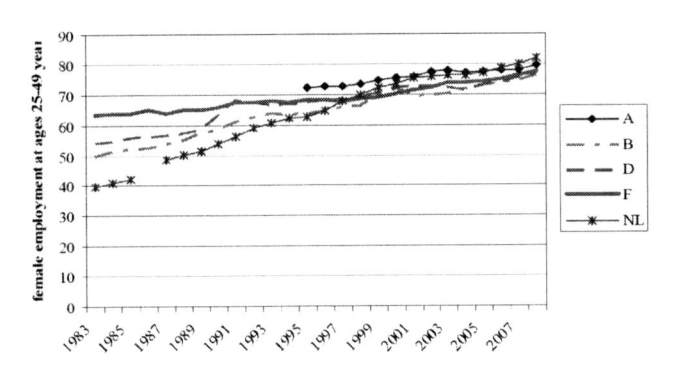

Spotlights on Contemporary Family Life - Chapter 3: Demographic Change and the Family in Europe

Western Europe

Southern Europe

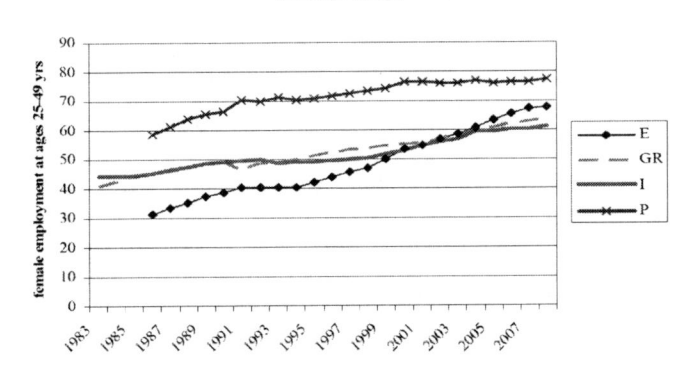

Central-Eastern Europe

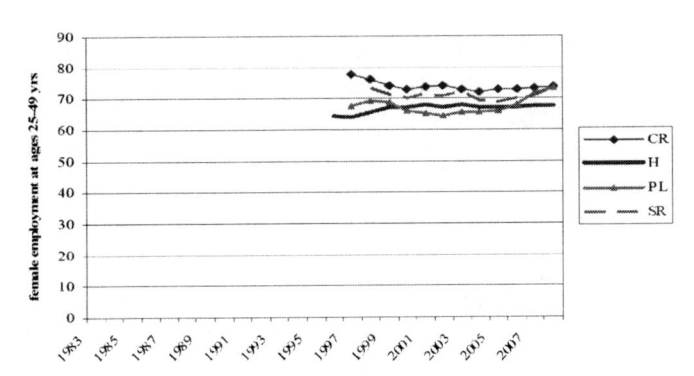

Source: *Eurostat [Employment and unemployment database (LFS)]*.

❖ *Economic theories have stated that low fertility is triggered by women's increased economic independence (see Becker, 1991). What is the so called "positive turn" described by Castles (2003)?*

Given the decline in birth rates and marriage rates, and a nearly simultaneous growth of female labour force participation in the developed world, women's increasing economic independence has been seen as a main cause of low fertility in economic theorising. Indeed, in the 1960s and 1970s, countries with modest levels of female employment showed high fertility rates, while societies with high levels of female labour market participation displayed low fertility. By the late 1980s however, the relationship between women's paid work and fertility changed. Since then, countries with high female employment rates have been the ones with fertility relatively close to the replacement level, while societies with women's more modest engagement in paid work have exhibited low fertility rates, often below the critical level of 1.5 children per woman. This is the so-called 'positive turn', underpinned by the decline in men's hourly wages and labour-force activity in the light of growing labour-market uncertainties and young people's difficulties in finding a stable job. Hence, women's employment has become almost necessary for couples starting a family.

❖ *Could you explain why, in some countries, like Sweden, fertility declined only moderately even though women engaged in paid work (70.2 per cent in 2009), while fertility rates below the critical level have been seen in other societies, like in Hungary, along with relatively low levels of female employment rate (49.9 per cent in 2009)?*

The *gender equity theory* (McDonald, 2000) in combination with *risk aversion theory* (Beck, 1999) provides us with insights into the rationale for this development. As pointed out by both theories, women and men in modern societies enjoy, for the most part, equal access to education at all levels and work for pay as individuals, irrespective of gender. The relatively high level of gender equity attained in these individual-oriented social institutions has not been matched in families in which women continue to perform the lion's share of household and childcare work. This in turn constrains their opportunities in other spheres, including the labour market, even if they are the main earner in the family, which is a less unlikely event in times of high unemployment and growing economic uncertainty. Having only one or two children, or maybe none at all, has thus become a strategy followed by women who wish to keep their options open, and/or to ensure a reasonable living standard for their families and for themselves in any case. Also, they can rely on modern, efficient contraceptives. Thus in societies where a large proportion of women see no other way than severely limiting their family sizes, fertility has fallen below the critical level.

❖ *In contrast to the mid-twentieth century, labour-force participation remains important for women even after they enter into a partnership or marriage and have children. How could you explain this change?*

High youth unemployment rates over an extended period in a number of European countries, combined with high economic aspirations and a reluctance to accept, if only temporarily, a lower living standard than in one's parental home, and the growing instability of couple relationships have strengthened the sense of being able to support oneself, irrespective of gender, among young people. Moreover, women are increasingly aware of the gender-unequal outcomes of partnership dissolution, which often leave women, much more so than men, to cope with economic hardship, especially if they have children. Hence young people, especially young women in modern societies, seek to minimise the risk of economic insecurity first and foremost by investing in their human capital, both in terms of educational attainment and employment experience. Irreversible decisions such as childbearing are carefully considered with respect to the timing and number of children one will have. The realisation of these plans is greatly influenced by the institutional context being perceived as supportive to - or a constraint on - the combination of parenthood and gainful employment. In the latter case, fertility can be locked below the critical level indefinitely.

❖ *The decline of fertility rates in Europe has been accompanied by a rising mean age of first time mothers. From an economic point of view, first parenthood at higher ages has many advantages for women. What are the risks of the postponement of first birth?*

The deferment of first birth is a high-risk strategy from the demographic and medical points of view. Fecundity declines with age, raising the need for assisted reproduction, which is costly and is associated with health risks for both the mother and the child (Gustafsson, 2001). It is also likely to lead to increasing levels of childlessness in a society, given both the biological thresholds and the socially accepted age limit for becoming a mother, but also the fact that people get accustomed to a childless lifestyle and may be increasingly unwilling to give up other priorities for the sake of parenthood. Rising childlessness levels increase the dependency burden over cohorts unless they are counterbalanced by larger family sizes among families with children, which is unlikely to occur in modern Europe. As smaller and smaller generations will have to support much larger parental generations, the need to reduce pensions as well as the quality and availability of public health care and elderly care will arise, making fertility a highly relevant issue for the future of the welfare state.

- ❖ *High levels of voluntary childlessness are a relatively new social phenomenon in Europe. In terms of the women who end up involuntarily childless, what are kinds of numbers are we talking about in Europe?*

It is of course difficult to be certain about how many of the childless in a country have never become parents by choice or due to medical reasons. In addition, involuntary childlessness may be seen by the individual herself to also include lack of (institutional or policy) support to reconcile parenthood and paid work, leading to the decision never to have a child.

On average, 3-6 per cent of a cohort is likely to have no children due to (reproductive) health problems. The proportion childless among younger cohorts is, however, much higher than that in most European countries. Among women born in the mid-1960s, nearly one-third in (West) Germany, one-fifth in Austria, Italy, the UK and (somewhat surprisingly) Finland, ended their reproductive years without ever having a child (Figure 5), while less than 13 per cent remained childless in France and in most Scandinavian countries, where achieving work-life balance is less of a problem than elsewhere. Central-Eastern European countries also displayed low levels of childlessness, given their pattern of early childbearing before the mid-1990s, which applied to those born up until the early to mid-1960s but has changed since then. Thus, a substantial increase in childlessness among women born after 1970 can also be expected in these societies, due to the difficulties of combining paid work and parenthood in the new market economies.

Figure 5: Proportion childless for female birth cohorts in 19 European countries

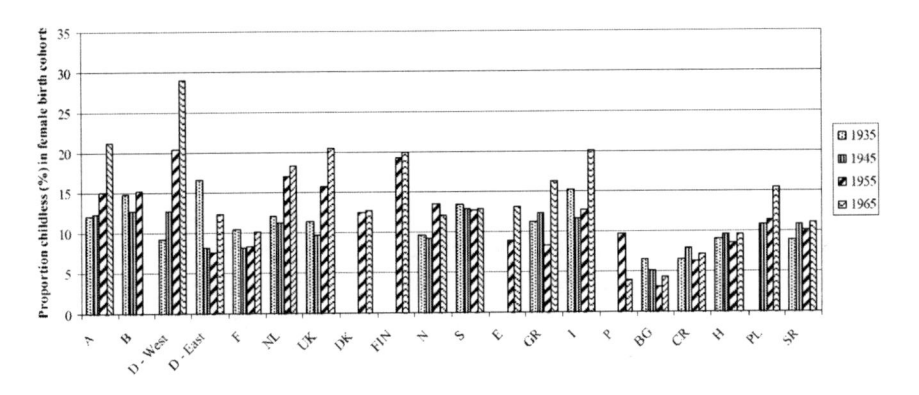

Source: OECD Family database; Sardon/Robertson (2004).

❖ *Since childlessness is not the main cause of declining fertility rates, it requires a separate explanation. Is there any evidence of factors that might explain the differences across countries by the proportion of women and men who end their reproductive years without ever having a child?*

Little is known about the level of childlessness among men. Based on data from countries where such information is available (e.g. Scandinavian societies), we may assume that the level of childlessness is even higher among men than among women. With respect to ideal family size, the latest Eurobarometer data addressing this topic (2006) showed that the share of women who prefer to have no children at all is rather small (slightly above 10 per cent, with the highest level seen in Austria). The discrepancy between preferred and realised number of children seems to be related to difficulties in reconciling employment and childrearing responsibilities (OECD, 2007). Indeed, combining information from Figure 5 with that on benefit and service provision to families in OECD countries, we find that childlessness is lowest in countries where high-quality public childcare is provided, with generous opening hours facilitating parents' labour market participation, to substantial proportions of children below age three, namely most Nordic countries and France[1]. But policy support is not the only factor that matters. Workplace cultures, in which long working hours and frequent overtime

---
[1] As already mentioned, we need to disregard the former socialist states due to their early childbearing patterns until quite recently, suppressing childlessness levels for the cohorts displayed.

hours are necessary conditions of career advancement - quite common in German- speaking countries and the UK especially - also appear to severely constrain women (and most probably men) in realising their childbearing aspirations.

❖ *Gender equality is often mentioned as a driver for slowing population growth. Fertility rates typically decrease as women acquire more control over decisions that affect their lives. What do demographic findings show? Does gender equality matter for fertility?*

The example of Sweden and other Nordic countries, where the principle of gender equality has influenced family policies (and public policies in general) for decades, and where fertility rates have never declined below the critical level despite high female employment rates, shows that gender equality does not suppress fertility, rather it can contribute to keeping it relatively close to the replacement level.

Indeed, empirical research has shown that the propensity to have a second or even successive children is higher in couples where the partners engage more equally in both earning and caring tasks than for other couples not only in Scandinavia, but also in Australia, New Zealand, USA, Spain, Italy and Hungary[2]. Hence, facilitating the combination of paid work and family responsibilities for both women and men can contribute to increase and/or keep fertility above the critical level.

❖ *It is controversially discussed whether governments should push for more gender equality in order to raise fertility. Family advocates plead for more focus on family policies and less on gender-equality policies. What are your main thoughts on the gender equality fertility proposals?*

Family policies and gender equality policies are not necessarily two separate entities, but can be closely interwoven and constructed to mutually support each other, as in Sweden. In fact, Swedish family policy does not aim to encourage childbearing, or to keep fertility at a certain level. It is the stated goal of Swedish gender equality policy to enable women and men to combine paid work with parenthood.

The main principle of social security is that every adult is responsible for his or her maintenance through own earnings unless incapable of paid work. In-line with this, labour force participation is encouraged, in-

---

[2] See Goldsheider et al, (2010) for an overview of studies.

dependently of gender and marital status, by individual taxation and lack of spousal alimony at divorce. Yet the level of child poverty in Sweden is amongst the lowest, and the level of maternal employment is amongst the highest seen in OECD countries. At the same time policy measures such as the income-related parental leave program and subsidised, high-quality public childcare facilitate the reconciliation of paid work and family tasks for both women and men, and men are actively encouraged to take part in childrearing and household duties. Besides specific features in the parental leave scheme, its flexibility and the high benefit level (currently covering 80 per cent, but before 1995 90 per cent of the previous income for the parent on leave), fathers' engagement in their children has also been strengthened by couples retaining joint custody for children after divorce or separation (since the early 1980s).

Such support for women's and men's equal engagement in paid work and family responsibilities has resulted in a feeling of security in their roles as earners and carers, which has contributed to Sweden exhibiting both high fertility rates and high employment rates among both women and men. Hence, this seems to be a successful strategy, worthy of consideration also by other countries.

❖ *Sweden was the first country in the world to convert maternity leave into a gender-neutral parental leave, in 1974. How has the take-up rate among fathers developed? What are the most important lessons the Swedish experience can teach other countries that experience low fertility levels?*

Indeed, Sweden was the first country in the world acknowledging fathers as caring parents on a par with mothers, in 1974. Fathers' role as care-providers has been further strengthened over time. Since 1980, fathers have been entitled to ten days' additional leave around a birth (so-called "daddy days"), paid at the same level as parental leave. In 1995, this was supplemented by a father's quota, which was introduced in the parental leave scheme and one month was reserved for the father (and another month for the mother), not transferable to the other parent. The quota was extended to two months in 2002, when the length of income-related parental leave was also increased by another month (to 13 months, plus 3 months covered with a low flat-rate benefit). Since June 2008, the program has also included a gender equality bonus to promote a more equal share of parental leave among mothers and fathers. As a result of this consequent policy development, fathers in Sweden increasingly engage in active parenting (Figure 6).

Figure 6: Uptake of parental leave among fathers in Sweden, 1986-2009

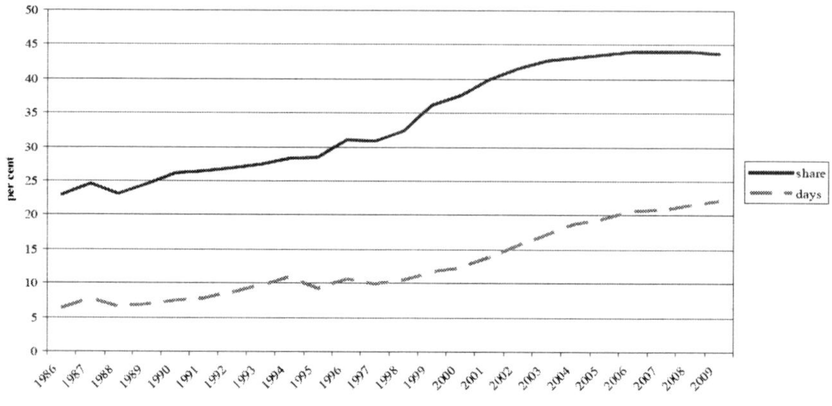

Source: *National Social Insurance Board (Sweden).*

Fathers' share amongst parental leave users has risen substantially since the mid-1980s, from about 25 to 45 per cent. The father's quota intensified the increase, even with respect to fathers' share of parental leave days, which increased from 10 to over 20 per cent over the last ten years. This also has a positive effect on fertility. Research in Sweden and Norway has shown that the propensity to have a second and third child is higher if the father took some parental leave with the previous child than in families where he used no parental leave at all (Oláh, 2003; Duvander *et al.*, 2010). This suggests that parents' more equal engagement in the care of their children is likely to contribute to the positive development of fertility in a country.

❖ *Recently, more and more demographic studies are focusing on the role of social policies and institutions in shaping fertility. In the article entitled "Birthstrikes? Agency and capabilities in the reconciliation of employment and family" (Hobson/Oláh, 2006) you pointed out the disjuncture between norms/values and practices and between policies and women's' capabilities of exercising them. What was the purpose of using the concept of birthstriking?*

In this article we sought to create a bridge across demographic and welfare state research, considering individual-level fertility behaviour in various institutional contexts. By using the concept of birthstriking, inspired by Amartya Sen's ideas on agency and capabilities shaping individuals' real freedom to choose, we studied the impact of different educational levels on women being able to combine employment with childbearing in four policy configuration models. Linking the individual level, i.e. women's resources and aspirations, with societal and institutional levels, based on data from the 1990s - a decade with fertility rates at unprecedentedly low levels in certain countries of Europe inspiring the concept of "lowest low fertility", when ideologies and policies for male breadwinning were also on the wane - we aimed to illuminate which individuals and families are unable to achieve what is increasingly considered a norm, i.e. combining paid work and parenthood, and are therefore delaying and/or not having children.

❖ *Could you tell us more about the results?*

Our results suggest that institutional settings seem to matter for childbearing decisions: societies with weak reconciliation policies and weak protection for workers are also societies where family formation is greatly constrained. Southern European countries, where unemployment rates, especially among women, have been very high and policy support for women to reconcile paid work and family responsibilities has been quite limited, are clear examples. Also societies with dramatic changes in the ability to achieve work-life (or work-family) balance and increasing economic uncertainties, as in Central-Eastern Europe, provide fertile ground for birthstriking.

❖ *According to the studies and research you carried out, could you identify some priorities for future research, which appear promising and could help us to better understand the mechanisms in the interplay between gender relations, the institutional context and individual/couple decision-making on childbearing?*

Indeed, a better understanding of the mechanisms in the interplay between gender relations, the institutional context and individual/couple decision-making on childbearing is of high policy relevance, as demographic sustainability is one of the key challenges Europe is facing. As the issues to be addressed are rather complex, research needs to be based on a comprehensive approach linking the individual, workplace, societal, and institutional levels.

An example of such an approach is the agency and capabilities approach, focusing on individuals' real freedom to choose. It offers a framework for addressing how diverse institutional contexts (including family policies, labour market structures and workplace culture) influence gendered capabilities and family formation, both intentions and their realisation, through facilitating the combination of labour force participation and childrearing for women and men, and by mitigating the costs of children for families.

In addition, we need to have better insight into the ways gender attitudes influence family career choices during the two stages of gender role change, i.e. female employment rates approaching those of men in the first phase, and male participation in home responsibilities becoming substantial in the second stage. This may require us to develop separate measures of public sphere gender equality and private sphere gender equality, to be taken into account in future large- scale surveys.

References

- Beck, U. (1999). *World risk society.* Cambridge, United Kingdom: Polity Press.
- Becker, G. S. (1991). *A Treatise on the Family.* Cambridge: Harvard University Press.
- Castles, F. G. (2003). *The world turned upside down: below replacement fertility, changing preferences and family-friendly public policy in 21 OECD countries."* Journal of European Social Policy 13.3: 209–227.
- Duvander, A. Z., Lappegård, T. & Andersson, G. (2010). *Family policy and fertility: Fathers' and mothers' use of parental leave and continued childbearing in Norway and Sweden.* Journal of European Social Policy 20.1: 45-57.
- Gustafsson, S. (2001). *Optimal age at motherhood. Theoretical and empirical considerations on postponement of maternity in Europe.* Journal of Population Economics 14.2: 225-247.
- Goldscheider, F., Oláh, L. Sz. & Puur, A. (2010). *Reconciling studies of men's gender attitudes and fertility: Response to Westoff and Higgins.* Demographic Research 22, article 8: 189-198. Available from: *http://www.demographic-research.org/.*
- Hobson, B. and Oláh, L. Sz. (2006). *Birthstrikes? Agency and capabilities in the reconciliation of employment and family.* Marriage and Family Review 39. 3/4: 197-227.
- McDonald, P. (2000). *Gender Equity, Social Institutions and the Future of Fertility.* Journal of Population Research 17.1: 1-16.
- OECD. (2007). *Babies and bosses – Reconciling Work and Family Life: A Synthesis of Findings for OECD.* Available from: *http://www.oecd.org/document/45/0,3343,en_2649_201185_39651501_1_1_1_1,00.html.*
- Oláh, L. Sz. (2003). *Gendering fertility: Second births in Sweden and Hungary.* Population Research and Policy Review 22.2: 171-200.

3.5 "To be or not to be and how to fill empty cradles? That is the question"

Zsuzsanna Kormosné-Debreceni
National Association of Large Families, Hungary

Hungary is one of the European countries that has been most affected by the demographic winter. This is due to the seemingly unstoppable decrease in birth rates, which started in the 1960s, and due to negative population growth since 1981. It is clear that all actors in society must co-operate not only to help slow this process but also to turn the corner and move towards a growth trajectory. The alternative is the wrecking of our systems of social security, education, and health care.

Hamlet's question is more pertinent than ever: the long lines of empty cradles throughout Europe and Hungary make us think of the possible answers to this eternal question and seek suitable responses.

As a representative of a family NGO, I do not seek to repeat research data and to reproduce the reasoning of academic experts. Instead I intend to show how we see this issue and how civil organisations can act. In addition, studies analysing the role, the possibilities, the experiences and the results of NGOs in supporting families should be interesting and useful to readers.

To be or not to be... untroubled? Can we ignore the fact that the Hungarian fertility rate is about 1.3, and most of European countries share the same or a similarly low rate?

Demographic experts say that about 80 per cent of Hungarian young people declare marriage, family and children as their main source of happiness and that they express a wish to have a number of children which would allow for some growth in the population. In reality, however, young people do not marry, or do so only very late. They have children later and later, and as a consequence have only one child or a maximum of two – a figure which is too low to maintain the size of population and thus society. The causes of this gap between desires (considered by some researchers only as formal responses to the supposed moral expectations of society) and reality have to be investigated more deeply. Several pieces of research by the Demographic Research Institute of the Hungarian Central Statistical Office and of the Institute of Behavioural Sciences of the Semmelweis University have studied possible causes, which have ranged from family values, traditions and models, to mental and spiritual obstacles, and from social and family policies and career and employment issues, to questions relating to the educational level of couples. A main causal factor identified by both European and Hungarian researchers is the tension between work career and private (family) life. In this article I will touch on some aspects of this element in the long list of causes.

Not only researchers, but the everyday experiences of families with children show that work-family tension is a real difficulty to be resolved. The problem is even harder in the case of families with several children, because employers avoid hiring these parents on the assumption that they will not be sufficiently reliable, because of problems caused by their large families. This is why the National Association of Large Families (NALF) in Hungary began to work in the field of reconciling work and family in 1999 and to cooperate with governments, participating in European and domestic forums, supporting the family-friendly workplace movement, talking to experts and promoting the issue in every possible way.

What have been the main areas of activity of NALF in this field during the past ten years? From 1999, as a member of different consultative bodies, we have been constantly repeating the fact that the problem of inequality in the labour market is much deeper in the case of parents (mainly mothers) with children than that of men and women in a general sense. We were part of the team defining the criteria of the 'Family-friendly Workplace Prize'[1] and of the expert group working on the adaptation of a related audit system[2]. Every year, we also award a special prize to an employer whose treatment of employees fits most closely with our values – for example by giving bonuses or provided support to employees with large families. After some years of regression of the movement due to a lack of a real political will, the new government seems to have the intention of reviving the movement and encouraging and motivating employers in different ways (e.g. by reducing their tax burden when employing parents part-time, through job-sharing, or by giving them preference in tenders, renewing contracts, etc.).

The year 2004, when Hungary became a full member of the EU, was very interesting. In March, the European Parliament accepted a Resolution named Work, the family and private life (P5_TA(2004)0152)[3] based on a petition signed by 63 family NGOs from 14 countries (among them our association). The Resolution pointed out that *"family policy should create conditions which enable parents to spend more time with their children and that in many cases a more equal division of parents' time between paid work and caring for their children would lead to better contact between parents and children and also have a positive effect on family formation and family stability and considers that a general reduction of daily working time is the best way of combining work and family life"* (point 3).

---

[1] "Családbarát Munkahely Díj" *(http://palyazatok.org/csaladbarat-munkahely-2010-dij/)*.
[2] Both the Movement and the Audit were taken over from Germany. For further information (in Hungarian) see *http://www.fiona.org.hu/downloads/ferfibeszed_zarotanulmany.doc*.
[3] See *http://www.europarl.europa.eu/sides/getDoc.do?type=TA&reference=P5-TA-20040152&language=EN*.

At the end of the same year a peer review of the work-family issue was organised in Berlin, and a delegate of NALF was invited to the meeting as an independent national expert. At this meeting, Germany presented its new initiative: the *Local Alliances for Family* (Lokale Bündnisse für Familie) as a new way of creating work-life balance and improving the wellbeing of families with children and, as a result, encouraging couples to have children and thus increase the fertility rate. This initiative met the aims and strivings of NALF and therefore we began to work on implementing this in Hungary. We have organised international conferences, published a booklet presenting the model and highlighting some Hungarian examples of family-friendly municipalities. In 2005 we launched the Family-friendly Municipality Prize and since then we have seen a slow but steady expansion of family-friendly measures in local government. In Germany the number of Local Alliances exceeds 600. In Hungary we are still at the beginning, with fewer than 10 officially created Alliances, but we do have 23 winners of the Family-friendly Municipality Prize. We are convinced that - besides a firm political will and different state measures and support pushing forward the issue of work-life balance - acting locally and involving all possible local actors is one of the most important tools for creating the basis of a better work-family balance, better family life, and thus for encouraging young couples to give birth to children.

One of the key points of the Local Alliances for Families is the co-operation of local employers, companies and entrepreneurs. We need to make them understand that being family-friendly is not a loss for them, but that they may even profit from employing parents who are usually loyal and grateful employees. In addition, co-operating with other local actors gives a very positive image to businesses – and this is also a good position to be in. They should be aware of the new skills which mothers acquire during the childcare period. Mothers' ability to divide listening, to keep to a budget, to work in a team, to discuss issues in such a way as to get to a win-win position, to communicate with persons of very different ages, educational levels and background, to negotiate with institutions and authorities, to organise events, etc., are skills which can also be useful in their work. The absence of employers and companies as stakeholders at the Lisbon FAMILYPLATFORM conference (despite some representatives being invited) and the lack of studies focussing on the employers' perspective give us the feeling of incompleteness. Without knowing their success or difficulties and their views related to the employment of parents we cannot find suitable solutions for balancing labour and family duties.

A report of the American *Sloan Work-family Policy Network* (Integrating Work and Family Life – a Holistic Approach, published in 2001) says that *"... jobs are still designed as if workers have no family responsibilities. The culture and organization of paid work, domestic care work, and community organizations remain predicated on the breadwinner-homemaker model. Thus, jobs, schools, medical services, and many other aspects of contemporary life operate on the assumption that someone (a wife) is available during the typical workday to care for children after school, during the summer, or on snow days, to take family members to the doctor or the dog to the vet, or to have the refrigerator fixed. And increasingly, the sisters, mothers, grandmothers, friends, and neighbors that working women (married or single mothers) relied on in the past, are themselves now in the labour force and, in the case of relatives, frequently live in another city. The new global economy, with its focus on 24/7 availability and long work hours, only worsens the problems generated by the lag in the organization of paid work, as if workers where without personal interests or domestic care concerns"* (p1-2). As a solution, the Report calls different actors to action (just as the model of Local Alliances does) saying: *"Employers, unions, professional associations and advocacy groups, government, and communities all have roles to play in integrating work and family life, but none of them can solve this problem acting alone"* (p2).

To be or not to be… a working mother leaving small children in institutions or in the care of persons who are not family members? There is a constant debate about the effects that a mother's early return to work has on her children. Experts' convictions are contradictory: some of them claim to prove that institutional care is far from being acceptable in the case of children under the age of 3 or even 6 years, while others encourage young mothers and fathers to return to the labour market as soon as possible, and urge governments to provide childcare institutions to help parents to go to work. In the case of children under the age of 3, the Barcelona target is for 33 per cent to be placed in crèches and other childcare institutions . If it is true that institutional care is detrimental for children of this age, how can one insist on placing one-third of our small children in childcare institutions instead of giving them the possibility of staying at home with their mothers or fathers (disregarding here cases where there is an absolute need to go to work early, such as during financial difficulties or unexpected job opportunities) One can argue that crèches are not the same as institutional care for orphans or abandoned children, or for those taken out of their families for child protection purposes, and the evenings and weekends are enough for filling the "emotional tank" of children and assuring the affection and attachment necessary for the healthy development of the child. But we need to ask: why do we have so many children with behavioural, mental and learning difficulties? The number of children "with special needs" seems to be growing, and education professionals and teachers think hard about the causes and the ways of helping these children overcome their difficulties.

The other aspect of the issue is the frustration and stress affecting mothers due to the double burden, the permanent pricks of their conscience, intensified by comments of people outside the family saying they neglect their children, and the fear of the child becoming ill, and they then losing their jobs. Do mothers and children really need these burdens? Is it really worth it for society to force women to return to work too early? Is it not better for all of us to give a choice to the mothers (and fathers) and give children a chance to grow up in an emotionally and physically secure environment with their own parents (and siblings), and avoid the negative consequences and the stress of being uprooted day after day from their home, delivered like packages to strangers and unfamiliar circumstances in a place which is foreign to them?

The situation is even more complicated in the case of mothers of several children. The management of such families often requires double full-time work at home, so the ability to balance outside work and family is rather restricted (despite their creativity and good organising skills). It is important to find out how parents of large families organise their family lives, and how care systems meet their special needs. At the Lisbon FAMILYPLATFORM conference it was revealed that a majority of researchers and studies do not consider this type of family, and there are only very few analyses examining the real values in these families, their positive effects on children and parents and on society, and their special difficulties and needs. As a large family organisation we call the attention of researchers and decision-makers to this group of families. In Hungary the proportion of large families is about 14 per cent of households with children, but 28 per cent of all children live in these families. The physical and mental wellbeing of a major part of the future generation should not be neglected.

The World Movement of Mothers (MMM) report *"Realities of Mothers in Europe"* (based also on the opinion of 580 Hungarian mothers) states: *"The economic dimension of family life, while essential, may not be as influential on the life of the child as closeness or connection that the child feels to the parents"* (p14). In the same report we find diagrams showing that 25 per cent of mothers prefer to take full-time care of their family and another 64 per cent prefer some combination of part-time work and family-care duties (p23, Fig. 3.2). Eighty per cent of them wish to take care of children under the age of three at home. In large families the need to care for children at home is even stronger. NGOs like NALF or MMM have an obligation to amplify such voices and to make politicians and decision-makers change their thinking and consider that 32 per cent of the mothers' messages to politicians mention the work-life balance as their main concern, but not necessarily in the way politicians speak about it. Mothers (and fathers) would like to make the choice themselves, without any outside pressure. NALF intends to disseminate the results of the MMM study during the Hungarian Presidency of the EU in the first half of 2011.

In the draft of the EU initiative "Roadmap for Reconciliation between Work, Family and Private Life", planned for adoption in 2012, it is pointed out that *"Civil society organisations (social NGOs, family organisations) have also been consulted generally on the reconciliation package, but a more targeted consultation of such stakeholders (other than social partners) will be part of the studies"*. We hope this consultation will not only be conducted, but also that the opinion of family NGOs based on the experiences and needs of the families and their background will be considered and used to build relevant European and national policy responses.

I began my train of thought with the famous but depressing words of Hamlet, Prince of Denmark. At the end of this article let me cite from the last passage of a famous Hungarian dramatic poem, the Tragedy of Man by Imre Madách. The main characters are Adam, Eve, Lucifer, and the Lord. After Lucifer shows Adam over the history of mankind in a dream, Adam seems to lose all hope, fights with the Lord and wants to die. But suddenly Eve whispers something to him:

> *"I know*
> *How happy you will be, so hear,*
> *I'll whisper low, oh come, come near*
> *I carry your baby, Adam dear."*

And this is the point at which Adam's thinking changes. From this point on he won't struggle with the Lord or listen to Lucifer who tries to destroy him. He finds his peace, the future of mankind is in this little new life. The last words of the poem are spoken by the Lord:

> *"Man, I have spoken: strive on, trust, have faith!*
> *That is also the answer to our question…"*

Chapter 4: Volunteering in Families

Editorial

Anne-Claire de Liedekerke, Joan Stevens and Julie de Bergeyck
MMM Europe (Mouvement Mondial des Mères – Europe)

This final chapter of the *Spotlights on Contemporary Family Life* is dedicated to volunteering and family. Not only has the European Commission announced 2011 as being the European Year of Volunteering, but the United Nations has also marked 2011 to celebrate the tenth anniversary of the "International Year of Volunteers" in order to find new areas where volunteering can make a difference.

As representatives of mothers across Europe and the world, Mouvement Mondial des Mères (MMM) has operated since 1948 thanks to unpaid volunteers, like thousands of other non-governmental organisations in Europe and across the globe. From a recent survey[1] of mothers launched by MMM across Europe, more than half (or 56 per cent) of the 11,000 respondents reported that they volunteer (mainly for non-profit associations, schools and faith-based institutions). Surprisingly, the age of the mothers is not the major differentiating factor – instead, it is the number of children they have. Indeed, the more children they have, the more they tend to volunteer: 39 per cent of the responding mothers with one child volunteer, 48 per cent of mothers with two children, 59 per cent of mothers with three children, 72 per cent of mothers with four children, 77 per cent of mothers with five children. Can we deny how mothers (and fathers) play an important role for their children in modelling the example of helping and volunteering? In a talk given in April 2007, an MMM affiliate in Lebanon recalled how mothers living in refugee centres during the recent armed conflict there dealt with everyday power conflicts to maintain peace and security: *"Mothers model the example of service in the family...Just after the cease-fire, my daughter came with me to the South. Following that event, my daughter said in a TV interview: 'When I was very small, I observed that even if we lacked nearly everything, we always had enough to share with the neighbours'"*.

The objective of this chapter is not only to pay tribute to the nearly 100 million Europeans who volunteer their time and talent to a cause and to families who instil values of volunteering in their children, but also to give voice to six contributors (from different professional backgrounds and dif-

---

[1] The European Survey of Mothers (Mouvement Mondial des Mères, 2011).
See *http://www.mmmeurope.org/en/results-european-survey-mothers*.

ferent countries) who have accepted the invitation to write an article on this important topic. These authors walk us through different definitions and concepts of volunteering, and shed light on the national differences and public policies (or lack thereof) in the European Union Member States. Two articles are dedicated to explaining the origins, aims, programmes, activities and future of the 2011 European Year of Volunteering from two different perspectives, the European Commission and Civil Society. Christiane Dienel explains how demographic change impacts on volunteer work, and Francesco Belletii and Lorenza Rebuzzini focus on how families educate to take voluntary action in Italy. Finally, we have the pleasure of including an article on volunteering in the US by Barb Quaintance.

All these articles demonstrate that families are a main source of voluntary help and assistance, and how volunteering has an impact on the lives of families in Europe. These articles also illuminate the need to strengthen institutional frameworks that support volunteerism, and highlight essential research gaps and questions.

To close with UN Secretary-General Ban Ki-Moon's message on Dec 5, 2010: *"Let us honour volunteering as an expression of our common humanity and a way to promote mutual respect, solidarity and reciprocity. It is a powerful means of mobilising all segments of society as active partners in building a better world."*

4.1 Volunteering in the European Union: An Overview of National Differences in the EU Member States

Birgit Sittermann
Observatory for Sociopolitical Developments in Europe
(Institute for Social Work and Social Education, Frankfurt)

The coach of the boys' football team in the local sports club, the women visiting patients in the hospital, or the pensioner who explains the history of a city's church to visitors – all have something in common: they are volunteers. We find them everywhere: in London and Lisbon, in Athens and Amsterdam. The European Year of Volunteering for the Promotion of Active Citizenship 2011 brings these volunteers into the limelight. According to a recent Eurobarometer survey, 30 per cent of all Europeans declare that they volunteer in an organisation or are participating actively in an organisation (European Commission, 2010: 171). A closer look at the data classified by country reveals great differences: in the Netherlands, Sweden and Denmark more than 50 per cent declare that they are engaged in volunteer activities; in contrast, less than 20 per cent of the Portuguese and the Bulgarians identified themselves as volunteers.

But national differences go beyond 'raw numbers' of volunteers: in the European Union, different traditions and different definitions of volunteering can be identified. As a result we find different approaches in national policies on volunteering. Furthermore, this article highlights the relationship between volunteering and families. But before taking a closer look at these aspects, we have to clarify what we mean by "volunteering".

1. Different definitions and understandings of volunteering

This question is easier to ask than to answer. Apart from the English word 'volunteering', other languages use different terms for voluntary activities with different connotations. The Germans speak of *ehrenamtliches Engagement*: this describes for instance volunteering as the chairperson of the local sports club or being a member of the city council. Another common German term is *bürgerschaftliches Engagement* (civic engagement) which highlights the idea of volunteers as active citizens but refers at the same time to all kinds of voluntary activities. In French, two different terms are known: *bénévolat* and *volontariat*. In France, *volontariat* refers to voluntary services; *bénévolat* describes individual voluntary activities for the benefit of the society (and not for family and friends) (GHK, 2010: 52). In Belgium, both terms are synonymous, but the 2005 law on volunteering uses only the word *volontariat* (GHK, 2010: 51).

Besides the use of the different terms with their special connotations, one must consider the different kinds of activities that can be subsumed under 'volunteering': is donating money volunteering? What about watering the plants of your holidaying neighbour or shopping for your grandmother? Does being a member of an association qualify you as a volunteer? Does volunteering only take place in the framework of an organisation or project (i.e. formal voluntary activities) or do spontaneous or unorganised voluntary activities (i.e. informal or non-formal voluntary activities) count as well?

Neither academics nor practitioners have one clear-cut answer: a variety of definitions exist, suited to different (national) contexts and purposes. The *Study on Volunteering in the European Union* (GHK, 2010: 49ff.) provides a good overview of the different definitions used by international organisations and in the different European Member States[1]. A current definition of volunteering can be found in the Decision of the European Council Decision on the European Year of Voluntary Activities Promoting Active Citizenship 2011[2]:

> "(…) the term 'voluntary activities' refers to all types of voluntary activity, whether formal, non-formal or informal which are undertaken of a person's own free will, choice and motivation, and is without concern for financial gain. They benefit the individual volunteer, communities and society as a whole"[3].

To agree on one concrete definition might not be necessary for every discussion, as the term 'voluntary activities' can refer to a broad range of activities. National differences have to be taken into account, however, especially in an international or European context. In Germany and France for example, informal volunteering is not included in the general understanding of volunteering. In contrast, in Austria informal volunteering, for example neighbourly help, is perceived as volunteering. The Swiss additionally include monetary and non-monetary donations in their concept of voluntary activities (Federal Ministry for Family Affairs, Senior Citizens, Women and Youth (2009)). The term volunteering in the Netherlands also refers to political participation and caring for young children and (elderly) relatives (Vogelwiesche/Sporket, 2008: 11).

---

[1] Funded by the European Commission, this study provides a good overview of the situation of volunteering in the European Union. It consists of country reports for every EU member state, a comparative summary of all results, and special reports on volunteering in the field of sport in the European Union. The study is available online at: *http://ec.europa.eu/citizenship/news/news1015_en.htm*.
[2] See *http://eur-lex.europa.eu/LexUriServ/LexUriServ.do?uri=OJ:L:2010:017:0043:0049:EN:PDF*.
[3] The decision does not make clear why both non-formal or informal voluntary activities are mentioned or what the difference is between these two terms. It is more common to use non-formal and informal voluntary activities as synonyms.

Special circumstances have to be considered for the post-communist EU Member States in Central and Eastern Europe. New voluntary associations had to be founded after democratisation. The citizens of these countries had to develop a new attitude towards volunteering because during Communist rule membership in youth organisations or participation in political festivities or demonstrations was mandatory rather than really voluntary (Zimmer/Priller, 2004; GHK, 2010: 48).

2. How many volunteers are there and what do they do?

The Study on Volunteering in the EU (GHK, 2010) pooled national studies on volunteering to analyse how many people volunteer in the EU. However, due to different methods and definitions of volunteering applied in the national surveys it is not possible to arrive at a precise number. The authors of the study concluded that 92 to 94 million adults in the EU are volunteers. That means that 22-23 per cent of all EU citizens aged over 15 are involved in voluntary work[4].

According to the Study on Volunteering in the EU, countries with a very high level of volunteering are Austria, the Netherlands, Sweden and the UK. In contrast, less than 10 per cent of adults volunteer in Bulgaria, Greece and Lithuania (GHK, 2010: 5ff.)[5]. In order to assess these different levels of volunteering it should be taken into account that in some countries, such as Greece, it is more common to dedicate one's free time to helping families or friends rather than to be a formal volunteer within an organisation. But most surveys on volunteering focus on formal volunteering in the framework of an organisation or a project (Angermann/ Sittermann, 2010: 10).

The most common fields European volunteers are engaged in are "sport/recreation/leisure", "culture and arts", "education and research", "social activities/social services", and "health". Again there are national differences, for example in Lithuania over half of the voluntary organisations (55 per cent) are active in the social service and healthcare sector. In Bulgaria, Ireland, Spain and Portugal, social services also account for the majority of volunteers. Sport is the sector with most volunteers in Belgium, Germany, Finland, France, the Netherlands and Latvia (GHK, 2010: 280ff.). What do volunteers do? The study identified six main fields of activity: "administrative and supporting tasks", "helping or working directly with people", "preparing and supporting voluntary activities", "managerial and coordination tasks", "campaigning and lobbying" and "organisation of events" (GHK 2010: 89).

[4] The discrepancy in relation to the Eurobarometer survey mentioned above can possibly be explained by the fact that the Eurobarometer study question asks about both volunteering and participation in an organisation. It would be desirable to conduct a comprehensive study on volunteering in the EU to finally have exact data on the level of volunteering in the EU.
[5] For a full overview of the situation in all EU countries, see GHK (2010: 60ff).

3. Volunteering by and for families

The activities of many volunteers revolve around families and children. Research on volunteering has rarely paid special attention to families who benefit from the voluntary work of others but at the same time are active as volunteers themselves. A German report published in 2009 looked more closely at volunteering by and for families. The authors analysed different studies and data available for the situation in Germany (*Wissenschaftszentrum Berlin für Sozialforschung*, 2009).

According to their data, 40 per cent of all volunteer work is dedicated to families or children and young people. Most volunteers working with children and families are active in the sports sector. Here, every other volunteer from the sports sector declared that he or she is engaged in working with families and young people. Other sectors in Germany characterised by volunteers working for families are "church/religion" (33.4 per cent of all volunteers in this field), "recreation and leisure" (29 per cent, e.g. accompanying children and youth travel tours) and "culture and music" (20.8 per cent, e.g. conducting a youth choir) (*Wissenschaftszentrum Berlin für Sozialforschung*, 2009: 100).

Besides benefiting from formal volunteering within organisation, families also benefit from informal volunteering. The extent of this support has not yet been quantified, according to the authors of the study. However, unpaid support by family members, neighbours and friends is a relevant resource for families, especially in terms of child-minding. Data from 2005 suggests that 13 per cent of working mothers and fathers rely on relatives, friends, and neighbours to mind their children. The results of the German Volunteer Survey (2005) demonstrate the extent of family networks: 78 per cent of all households with children aged up to seven can count on the support of relatives, 52 per cent on the support of friends, and 34 per cent on the support of neighbours. (*Wissenschaftszentrum Berlin für Sozialforschung*, 2009: 110).

Naturally, families contribute to these informal networks as well. But besides this informal engagement, they are also volunteers in more formal settings. In fact, figures from Germany show that the level of volunteering among adults living with children is significantly higher than the average level of volunteering, though this does not apply to single parents (*Wissenschaftszentrum Berlin für Sozialforschung*, 2009: 102). The voluntary activities of these parents are to a high degree linked to their own children: three-quarters of all volunteering women state that their volunteering is directly connected to their own children (*Wissenschaftszentrum Berlin für Sozialforschung*, 2009: 103). Their activities revolve around child care facilities, their children's schools or leisure time activities, such as sports clubs. Parents volunteer as members of the parents' council, or they organise festivities or contribute their handcraft skills (*Wissenschaftszentrum Berlin für Sozialforschung*, 2009: 118). In Germany, more women than men volunteer in the fields relevant for families and children (*Wissenschaftszentrum Berlin für Sozialforschung*, 2009: 144).

As volunteers, parents can become role models for their children. Though not yet well researched, the few existing sources on this suggest that children imitate their parents' attitude towards volunteering and eventually become volunteers themselves (*Wissenschaftszentrum Berlin für Sozialforschung*, 2009: 102).

4. Public policies on volunteering

Different national traditions are reflected in the way public policies on volunteering have developed. Volunteering is an established, though not always prominent field of policy in many countries. Spain and Belgium have, for example, laws that define volunteering; France, Sweden and England, on the other hand, intentionally have no special legal framework for volunteering. Responsibility for volunteering lies not just at the national level but at sub-national level as well, in Germany, Belgium and Spain, where all autonomous regions have their own laws on volunteering. In the United Kingdom, the governments of Wales and Scotland are in charge of volunteering, but there is no policy for the whole of the UK (Angermann/Sittermann, 2010; GHK, 2010: 10). In England, the responsibility for volunteering lies with the Office for Civil Society (part of the British Cabinet Office). England is an interesting example for public policy on volunteering, as the change of government in 2010 led to a change in the policy on volunteering. Whereas the former Labour government focused on the promotion of volunteering by engaging with existing large voluntary organisations, the new coalition government emphasises the promotion of grassroots movements at local level. Their stated aim is to enable communities to initiate their own volunteer groups and projects. Another new programme in England is the National Citizen Service, which will start in summer 2011. This will be a (non-mandatory) voluntary service for 16-year-olds, who will use their summer holidays to develop a social project in their local community and put it into practice (Sittermann, 2011).

Some countries in the European Union have developed special strategies for their policies on volunteering. In Germany for instance, the government adopted a national strategy on volunteering in October 2010, which is the basis for the further development of the national policy on volunteering

(Angermann/Sittermann, 2010: 2). In Spain, the Fourth State Plan on Volunteering was to be implemented in 2010. The Spanish state's plans bring together different actors such as policy makers, representatives of voluntary organisations and experts on volunteering who work for the further promotion of volunteering (Sittermann, 2011).

The aim of public policies on volunteering is in general the promotion and facilitation of volunteering. One issue for volunteering policy is to make sure that volunteers have health, accident and liability insurance. Despite national differences on volunteering, one aspect is of relevance to all countries: acknowledgment and recognition of volunteers and their work. One common means of doing this is issuing volunteers with bonus cards which give them certain benefits, such as free access to museums. Additionally, several awards have been created to acknowledge voluntary activities. Examples are the Europe for Citizens Programme Golden Star Awards or the British Queen's Award for Voluntary Service. These awards place volunteers in the limelight, but general acknowledgement of voluntary work should exist beyond these brief moments and beyond the European Year of Volunteering 2011. This cannot be achieved by public policies alone, but requires an effort from everyone: when was the last time you thanked a volunteer - maybe the volunteer who issues the books in your local library or the voluntary firemen and women in your home town?

References

- Angermann, A., & Sittermann, B. (2010). *Bürgerschaftliches Engagement in den Mitgliedsstaaten der Europäischen Union – Auswertung und Zusammenfassung aktueller Studien.* Arbeitspapier Nr. 5 der Beobachtungsstelle für gesellschaftspolitische Entwicklungen in Europa. Available from: *http://www.beobachtungsstelle-gesellschaftspolitik.de/veroeffentlichungen/andere-veroeffentlichungen.html.*
- European Commission (2010). *Eurobaromètre 73. Rapport Volume 1.* (Terrain: mai 2010, publication: novembre 2010). Available from: *http://ec.europa.eu/public_opinion/archives/eb/eb73/eb73_vol1_fr.pdf.*
- Council of the European Union. (2009). *Council decision of 27 November 2009 on the European Year of voluntary activities promoting active citizenship* (2011) (2010/37/EC).
- Federal Ministry for Family Affairs, Senior Citizens, Women and Youth. (2009). Monitor Voluntary Activities Issue No. 1: *National and international Status of Voluntary Activities Research.* Available from: *http://www.bmfsfj.de/RedaktionBMFSFJ/Broschuerenstelle/Pdf-Anlagen/monitor-engagement-engl,property=pdf,bereich=bmfsfj,sprache=de,rwb=true.pdf.*

- GHK. (2010). *Study on Volunteering in the European Union* – Country Report France. Available from: *http://ec.europa.eu/citizenship/news/news1015_en.htm*.
- Sittermann, B. (2011). V*olunteering in the European Union. Creating a supportive environment and attracting volunteers.* European Conference (Berlin, 11 and 12 November 2010). Conference report. *Available from: http://www.sociopolitical-observatory.eu.*
- Vogelwiesche, U. & Sporket, B. (2008). *Strategien zur Stärkung des bürgerschaftlichen Engagements älterer Menschen in Deutschland und den Niederlanden. Kurzexpertise im Auftrag des BMFSFJ.* Available from: *http://www.engagement-conference.info/resources/D-NL-Expertise.pdf.*
- Wissenschaftszentrum Berlin für Sozialforschung. (2009). *Bericht zur Lage und zu den Perspektiven des bürgerschaftlichen Engagements in Deutschland.* German Federal Ministry for Family Affairs, Senior Citizens, Women and Youth. Berlin. *Available from: http://www.bmfsfj.de/bmfsfj/generator/RedaktionBMFSFJ/Broschuerenstelle/Pdf-Anlagen/buergerschaftliches-engagement-bericht-wzb-pdf,property=pdf,bereich=bmfsfj,sprache=de,rwb=true.pdf.*
- Zimmer, A. & Eckhard, P. (eds.) (2004). *Future of Civil Society: Making Central European Nonprofit Organizations work.* VS Verlag für Sozialwissenschaften.

4.2 Volunteers in the EU Spotlight: The European Year of Volunteering 2011

John MacDonald and Sara Lesina
European Commission

The year 2011 has been designated the European Year of Volunteering (EYV) to highlight the contribution made by volunteers from all walks of life to our economy and society. Volunteering has moved into the limelight in recent years, and the EYV will be an occasion to celebrate the importance of volunteering in creating a more democratic, caring and responsible society. To highlight volunteers' work and to encourage others to join in and address the challenges they face, the 2011 European Year of Volunteering was conceived with four main objectives in mind.

1. To foster an enabling environment for volunteering in the EU. The Year will help bring to light existing legal, administrative or other obstacles to volunteering in the Member States. By fostering an exchange of good practice between the Member States, the Year will help to implement appropriate measures to remove the barriers that are identified.
2. To empower volunteer organisations and improve the quality of volunteering. The European Year will provide input for further policy development on volunteering issues within Member States, and will initiate a dialogue between the EU Member States and Europe's developing world partners on volunteering issues. The aim is to encourage co-operation, exchange and synergies between volunteer organisations and other sectors, such as the government and corporate sectors, at European, national and regional levels.
3. To reward and recognise volunteering activities. The Year will improve the validation and recognition of skills and competences that can be gained through volunteering.
4. To raise awareness of the value and importance of volunteering. The Year will ensure that there is heightened awareness both within Europe and in partner countries of the value of volunteering and its contribution to the economy, society and the individual.

1. Background: volunteering and EU policies, programmes and activities

There have been a number of political developments in the area of volunteering since 1997, when an intergovernmental conference adopted 'Declaration

38 on Volunteering'[1]. The Declaration, which was attached to the final act of the Treaty of Amsterdam, recognised the importance of the contribution of voluntary activities to developing social solidarity. The Declaration stated that a European dimension of voluntary organisations would be encouraged, with a particular emphasis on the exchange of information and experiences.

Following Declaration 38, other EU-level documents emphasised the role of volunteering and committed to supporting volunteers across Europe. With the 'Recommendation of the European Parliament and of the Council of 10 July 2001 on mobility within the Community for students, volunteers, and teachers'[2], the European Parliament intended to give more opportunities for studying, training and volunteering across Europe by dealing with barriers to mobility. The European Parliament also encouraged Member States to ensure that the specific nature of voluntary activity is taken into account in national legal and administrative measures, to promote recognition of voluntary activities through certificates, and to take measures to ensure that recognised voluntary activities are not treated as formal employment.

Over the last few years, the issue of the social and economic value of volunteering became a central focus of EU-level documents. In March 2008, the European Parliament adopted a report on the 'Role of volunteering in contributing to economic and social cohesion'[3] which encouraged Member States and regional and local authorities to recognise the value of volunteering in promoting social and economic cohesion. In this document the European Parliament called on Member States to produce regular satellite accounts[4] as a complement to their usual National Accounts so that the value of Volunteering and Not-for-Profit Institutions (NPIs) could be measured.

In July 2008 the European Parliament adopted a written Declaration calling for a *European Year of Volunteering in 2011*[5]. The proposal for a Council Decision on the European Year (2011) was subsequently adopted on 3 June 2009[6], with the formal legislative base for the Year adopted by the Council on 27 November 2009[7]. Launching the European Year, the European Commission intended to raise awareness of volunteer engagement in Europe and to enhance volunteer activities.

---

[1] Declaration 38 on voluntary service activities, *http://eurlex.europa.eu/en/treaties/dat/11997D/htm/11997D.html*.
[2] Recommendation of the European Parliament and of the Council of 10 July 2001 on mobility within the Community for students, persons undergoing training, volunteers, teachers and trainers (2001/613/EC).
[3] European Parliament resolution of 22 April 2008 on the role of volunteering in contributing to economic and social cohesion (2007/2149(INI)) *http://www.europarl.europa.eu/sides/getDoc.do?pubRef=-//EP//TEXT+REPORT+A6-2008-0070+0+DOC+XML+V0//EN*.
[4] Satellite accounts provide a framework linked to the central accounts and which enable attention to be focussed on a certain field or aspect of economic and social life in the context of national accounts; common examples are satellite accounts for the environment, or tourism, or unpaid household work.
[5] Written Declaration 0030/2008 of 15 July 2008.
[6] Brussels, 3.6.2009 COM(2009) 254 final 2009/0072 (CNS) Proposal for a COUNCIL DECISION on the European Year of Volunteering (2011) {SEC(2009)725}*http://ec.europa.eu/citizenship/pdf/doc828_en.pdf*.
[7] Council Decision 2010/37/EC on the European Year of Voluntary Activities Promoting Active Citizenship (2011). See *http://eur-lex.europa.eu/LexUriServ/LexUriServ.do?uri=OJ:L:2010:017:0043:0049:EN:PDF*.

2. Doing better by doing good: the added value of volunteering

Almost 100 million Europeans engage in voluntary activities and through them make a difference to our society. Volunteering plays an important role in sectors as varied and diverse as education, youth, culture, sport, environment, health, social care, consumer protection, humanitarian aid, development policy, research, equal opportunities and external relations. Volunteering matters because volunteers translate fundamental European values on promoting social cohesion, solidarity, and active participation into action every day.

Volunteering contributes to building a European identity that is rooted in these values. Volunteers gain mutual understanding of people; it is indispensable in a wide range of EU policy areas, such as social inclusion, lifelong learning opportunities for all, policies affecting young people, intergenerational dialogue, active ageing, integration of migrants, intercultural dialogue, civil protection, humanitarian aid, sustainable development and environmental protection, human rights, social service delivery, increasing employability, the promotion of an active European citizenship, fighting the "digital gap", and as an expression of corporate social responsibility.

Volunteering is economically important too: the voluntary sector is estimated to contribute up to 5 per cent of GDP to some Member States' economies. So, the European Year of Volunteering should be a celebration of the valuable contribution that these millions of citizens make every day to our economy and society. Volunteering is freely given, but it is not cost free – it needs and deserves targeted support from all stakeholders: volunteering organisations, government at all levels, businesses, and an enabling policy environment and volunteering infrastructure.

3. The volunteering landscape in Europe[8]

The European volunteering landscape is extremely varied because of different historical, political and cultural attitudes towards volunteering in each country. The figures below give a more precise idea of the situation of volunteering in the EU.

The total number of EU volunteers is estimated to be around 94 million adults, which corresponds to 23 per cent of all Europeans over 15 years of age. The statistics do suggest that there are big differences in the level of volunteering between the EU's member countries. Whilst certain EU Member States have longstanding traditions of volunteering and well developed voluntary sectors (such as Ireland, the Netherlands, and the UK), in others the

---

[8] Source: *European Commission-DG EAC, 2010. Volunteering in the European Union. Final report. London: GHK.*

voluntary sector is still emerging or poorly developed (for instance in Bulgaria, Greece, Latvia, Lithuania, and Romania). National studies on volunteering show that the level of volunteering is: very high in Austria, the Netherlands, Sweden and the UK, with over 40 per cent of adults in these countries involved in carrying out voluntary activities; high in Denmark, Finland, Germany and Luxembourg where 30-39 per cent of adults are involved in volunteering; medium-high in Estonia, France and Latvia, where 20-29 per cent of adults are engaged in voluntary activities; relatively low in Belgium, Cyprus, Czech Republic, Ireland, Malta, Poland, Portugal, Slovakia, Romania, Slovenia and Spain, where 10-19 per cent of adults carry out voluntary activities; and low in Bulgaria, Greece, Italy and Lithuania, where the statistics suggest that less than 10 per cent of adults are involved in voluntary activities.

However, any such apparent differences need to be treated with caution, because there is a lack of internationally comparable data on volunteering. Each country has a different definition of volunteering, and different ways to measure it, so it is extremely difficult to make international comparisons. That said, over the past ten years, there has been a perceptible increase in the number of active volunteers and voluntary organisations in the EU.

Survey data from the EU-wide Eurobarometer survey (see Figure 1 below) carried out in 2006 suggests that the percentage of citizens who declare that they actively participate in - or do voluntary work for - an organisation varied from 60 per cent in Austria (the highest level of participation) to 10 per cent in Bulgaria (the lowest) in 2006. Overall, it would appear that the countries with highest percentages of volunteers are western European countries (and Slovenia) with well developed and established voluntary sectors.

Figure 1: Extent of active participation or voluntary work in the EU (%), 2006, according to the Eurobarometer survey (European Social Reality)

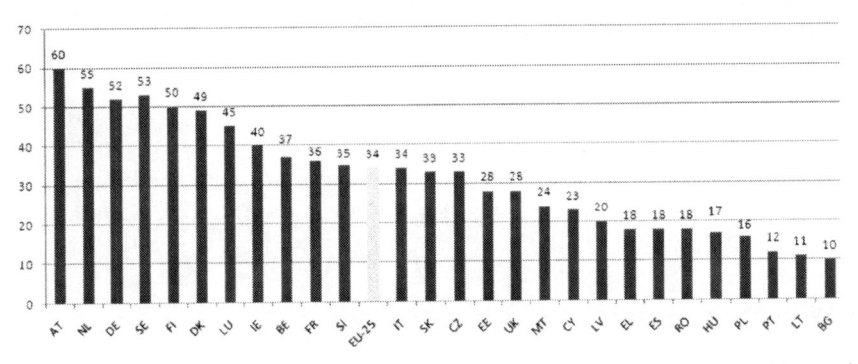

Source: http://ec.europa.eu/public_opinion/archives/ebs/ebs_273_en.pdf and GHK Consulting on the basis of Eurobarometer survey data (2009).

Moreover, in many countries a *gender dimension* is more apparent in specific sectors (e.g. sport, health, social and rescue services) and voluntary roles (e.g. managerial and operational roles) than in overall participation rates in volunteering. In general, most countries tend to have either more male than female volunteers (eleven countries) or an equal level of engagement (nine countries); this slight dominance of male volunteers overall can be explained by the fact the sports sector attracts the highest number of volunteers, and more men than women tend to volunteer in sport. For example, in Denmark there are important statistical variations between the participation of men and women in different areas of the voluntary sector. Men are considerably more involved in sports clubs and local community activities compared to women. At the same time women are significantly more involved in health and social service-related work than men.

Voluntary activities take place in many *different sectors*. According to the 2010 Eurobarometer survey, in over half of the EU countries most volunteers are active in the field of sports, exercise and outdoor activities (34%). Volunteers in sport represent an important share of total volunteers in Denmark (31.5%), France (25%) and Malta (84%). Other popular areas are social welfare and health (8%), charity and religious organisations (17%), cultural organisations, recreation and leisure, educational organisations, training and research (22%).

As for the landscape of *voluntary organisations*, there have been big increases in the number of voluntary organisations over the past decade: some countries have seen a two- or even fourfold increase in the number of registered voluntary organisations in the last decade, with individual annual increases reaching 15 per cent in some cases. These include countries where organised volunteering is an established tradition (like France and Germany), as well as countries where formal volunteering is a more recent phenomenon (such as Bulgaria, Estonia, Italy, Romania). However, it is important to remember that the level of detail on the number and sector of voluntary organisations depends on whether the country has a registry of voluntary organisations and whether such organisations are either encouraged or obliged to register. Even in countries which have such registries, it is difficult to provide accurate data on the number of active voluntary organisations, because in many cases the registries include both active and inactive organisations.

4. What goes wrong: obstacles and challenges in volunteering

Volunteering mirrors the diversity of European society: young and old, women and men, employees and unemployed, different ethnic groups and beliefs – all are involved in volunteering. However, seven in ten people do not volunteer, and in many cases this is because of real or perceived barriers to volunteering. These barriers take many different forms, such as a lack of information on how to become involved, time pressure, scarce economic resources, and the feeling of not being able to "afford" to volunteer. In some of the former communist countries, there is even a negative image of volunteering stemming from times when volunteering was compulsorily imposed. The challenges for volunteering vary from country to country depending on the national context. However, a number of common challenges across Europe can be highlighted.

There is a *lack of homogeneous data* on - and monitoring of - voluntary activities in EU Member States. As mentioned earlier, internationally comparable information and data relating to volunteering is rare and often unstructured and non-standardised, even at a national level. This clearly represents a major challenge in terms of accurately understanding volunteering within countries, in particular the impact of governmental support on volunteering in different European countries. This happens because of the absence of internationally comparable statistics and agreed methods of measurement, which could result in an unfair distribution of EU funds if different organisations in different countries measure voluntary contributions in different ways.

It is rare for there to be a *national volunteering strategy*: in total only five Member States have national strategies in place for volunteering – Austria, Estonia, Latvia, Poland and Spain. In countries that do not have a national strategy, the policy aims and objectives for volunteering are implicit within a wide range of broad policy discourses.

In many countries, there is a *lack of a dedicated legal framework for volunteering* to cover the rights and obligations of volunteers, such as the social insurance coverage of volunteers, and their training, entitlement to holidays, accommodation or 'pocket money'. Due to the many laws and regulations that non-profit organisations may have to follow, many NGOs are not aware of certain advantageous provisions. A clearer legal framework would clarify the position of paid and voluntary staff in voluntary organisations.

Moreover, there is an increase in the *professionalisation* of the volunteering sector, which is causing a growing mismatch between the needs of volunteering organisations and the aspirations of volunteers. Younger volunteers, for example, are less willing to commit to longer-term volunteering periods, even though this is increasingly requested by many volunteering organisations. The increasingly professional nature of personnel employed in the not-for-profit sector is also a challenge that voluntary organisations have to address: volunteers find themselves working side-by-side with newly employed paid professionals, recruited on the basis of specific skills.

Coupled with this is the problem of a *lack of recognition of skills* and competences gained through volunteering activities. Several European reports, such as the European Volunteer Centre's *Manifesto for Volunteering in Europe*[9] , have highlighted the lack of national systems promoting recognition in volunteering. Not enough research on the value of volunteering has been collected, and there is therefore insufficient recognition of its importance and insufficient 'evidence' on which to base or defend policy. In addition, there is little pan-European use of validation mechanisms for the skills and competences acquired through volunteering. Such mechanisms as do exist tend to be specific to one organisation (e.g. Red Cross) or programme (e.g. the European Voluntary Service), but there are increasing calls for more general validation or recognition mechanisms (such as a 'volunteer skills passport').

Sustainable funding is an increasingly pressing problem for volunteering organisations: there has been a big increase in volunteering organisations in recent years, and there is more competition amongst them for a shrinking pot of funds. The current austerity measures adopted by many governments have led to significant cutbacks in funding for volunteering NGOs. In addition, over the last few years there has also been a change in the way the public sector disburses public funds – away from grants and subsidies, and towards contracts awarded through public calls for tenders and a competitive bidding process. NGOs report greater difficulties in accessing funding through the latter channel.

5. "Volunteer – Make a difference!" – Touring Europe with a clear message: the EU would not be the same without volunteers

In dedicating the European Year 2011 to volunteering, the European Union is acknowledging the importance of volunteering in creating a more democratic, caring and responsible society. The European Year will raise awareness of the contributions of, and the challenges faced by millions of volunteers across the EU, whose efforts help to create a more democratic, caring and responsible society.

A good way to bring the European Year closer to the general public is through the interactive EYV 2011 'Tour'. The Tour will travel throughout 2011 to visit every Member State's capital city for a period of up to 14 days. It will provide volunteers and volunteering organisations with an opportunity to showcase their achievements, meet one another and discuss key issues for the future of volunteers.

[9] See *http://www.cev.be/64-cev_manifesto_for_volunteering_in_europe-en.html.*

Citizens and volunteers will also be able to find out about aspects of volunteering in other Member States and learn about the European dimension of volunteering. The Tour will provide a platform for understanding the world of volunteering and also add visibility to communication initiatives taken at the national, regional and local levels, attracting the media and public attention to the campaign.

The European Tour of Volunteering 2011 will feature an 'EU Corner', highlighting the European dimensions of volunteering, stories showcasing the experiences of volunteers, debates with policy makers, volunteers and citizens, meetings with volunteers, entertainment with family activities and intergenerational dialogue. The route started on 3 December 2010 in Brussels and will travel though all 27 EU Member States during 2011.

The EYV Relay

The EYV Tour will be accompanied by the EYV 'Relay'. The EYV Relay team is composed of 27 volunteer reporters, whose task is to report on the extraordinary stories of volunteers across the EU. The Relay Reporters, each coming from a different EU country, will depart from their home country to report for two weeks on the volunteers in another EU country. They will capture their experiences in film, sound and written articles.

The EYV EU-level thematic conferences

In order to push forward policy debate on volunteering issues at the EU level, the European Commission will organise a series of EU-level thematic conferences during 2011. Four conferences have been planned, each of them focusing on a different theme and target group, such as policy makers, volunteering organisations and the volunteers themselves. The conferences will promote a rich exchange and debate on important issues in volunteering between policy makers, businesses and volunteers, while attracting media interest. The first conference took place in Budapest on 8 January 2011, with the topic *"Recognising the contribution of volunteering to economy, society and the individual – where are we now, and where do we want to go?".*

Information about the Tour, the Relay and the EYV campaign in general will be provided on the official EYV 2011 website: http://www.europa.eu/volunteering. Materials about every step of the campaign will be uploaded. It will contain various sections giving information about the campaign and regular updates of its progress in the course of the year.

6. The year after: what will the legacy of EYV be?

The European Year of Volunteering 2011 is meant to help volunteering organisations and the volunteers themselves. The European Year will provide a much-needed impulse to set in motion the necessary changes, mainly at national level, that will make it easier for volunteering organisations and volunteers to do their work, and to do it better than ever before. Through a number of flagship initiatives, concrete measures will be taken to ensure people are better skilled and better prepared to face the challenges of the new economy. Therefore, the European Year of is not a 'one-off' year: it is the start of a process that will go well beyond 2011.

During the Year, and in the years thereafter, awareness will be raised about where change needs to occur, and these changes will be different in each country. The Year is a platform for broadening and deepening both the outreach and the quality of volunteering. The European Commission is working to ensure that volunteers all over Europe have been enabled - and continue to be so - to meet and learn what is done best in each European country. The awareness campaign of the EYV 2011 will help civil society and governments face these challenges, and start the work for necessary, beneficial change. Countries will examine where it may be necessary to revise laws or enact separate legislation in order to promote volunteering, protect volunteers and remove superfluous legal impediments.

4.3 Demographic Change and Its Impact on Voluntary Work

Christiane Dienel
Nexus Institute for Cooperation Management and Interdisciplinary Research

Demographic change is often discussed and presented as though it threatens social cohesion and the stability of the welfare state. How will the present level of welfare be maintained, when fewer children are born to be future contributors to welfare security systems and when longevity gives more years to nearly everybody? Under such circumstances, volunteer activity seems to offer an escape. Why couldn't all these active and healthy senior citizens help care for older members of society, often struggling with dementia? Couldn't civil society thrive, not in spite of, but because of demographic change?

The following article will give some background to this assumption and is mostly based on German data and a study done for the German Friedrich-Ebert-Stiftung (Dienel, 2010)[1].

European societies are ageing quickly. We will be "fewer, older and more colourful", but there is no consensus on whether this change will lead to a better society or to more problems. No doubt, dealing with demographic change is an enormous task, because it means providing service for the large number of "baby-boomers" born in the 1950s and 1960s who will retire from 2015 onwards. The base of contributors paying their pensions is narrowing. These processes, however, are developing rather slowly and leave us time to adapt. Regions in former socialist countries will face dramatic changes, though, because the significant fall in the birth rate there since 1989. But already the constant ageing of society diminishes the potential for volunteering, because the generation aged over 60 is less willing or able to participate than the younger members of the population. The graph shows the willingness for voluntary work (mid-grey) and the rate of actual engagement (dark grey) for the age groups 14-24, 25-59 and 60+ years in Germany.

1. Young people and volunteer activity

The participation of young people in voluntary activity seems essential in an ageing society. It is considered useful and instructive, it prevents violence and xenophobia, and helps to develop social skills. This kind of utilitarian argument is not very suitable for convincing young people to invest their time if they don't have fun while doing so. For young people, the key to motivation is not usefulness but the possibility of being independent, of developing their own ways of doing things, and of hav-

---

[1] See *http://library.fes.de/pdf-files/kug/07290.pdf*.

ing an impact on society. Self-determination of young people will not fit very easily into schemes made for them by those concerned about the future of the welfare state. Young people's civil engagement is often irritating and shocking, and seems destructive or disrespectful. It will not just fill the gaps left by the welfare state. More than two-thirds of German youth agree with the statement "I believe that politicians don't take young people seriously" and this feeling will be reinforced if they are only offered pseudo-participation arenas like youth parliaments without a budget.

Figure 1: Willingness to undertake voluntary work in different age groups, Germany

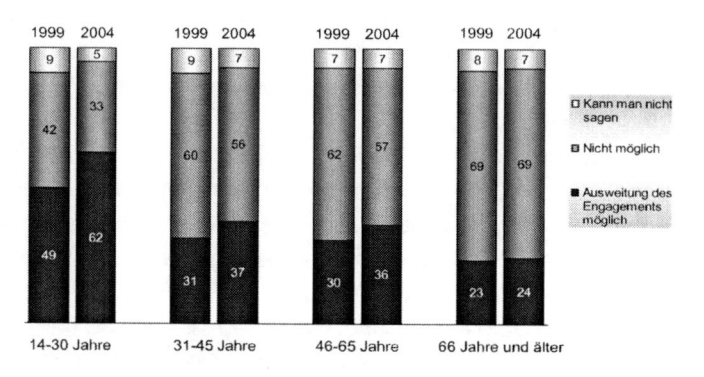

Source: *Gensicke/Picot/Geiss (2005: 213)*.

Still, many young people are very active in different forms of voluntary activity or civil associations. Sport is number one, followed by church and social activities. However, fewer young people are able to participate in time-consuming activities that require their presence once a week or more. Educational demands and preparation for professional careers often exert enormous stress, and simply don't leave enough spare time to invest in regular volunteer activity. Voluntary engagement is strongly correlated with level of education. But those young people who are successful in secondary education and who might be the group most interested in volunteering are also those who share the performance goals of their teachers and parents and will sacrifice other activities, if necessary, to achieve better grades.

In conclusion, it is no longer sufficient to develop attractive forms of volunteering for the participation of young people today. It is also indispensable to balance these offers with school schedules, work market pressure and biographical stresses. Demographic change not only means that there will be fewer children and young people to participate in voluntary activities, but also that those who are there have to face aspirations of parents and society and have less spare time than their mothers and fathers had. A solution would be to integrate voluntary activities into the school syllabus, to give certificates for certain types of engagement and, above all, to leave some free time and space to the younger generation.

2. The middle generation and voluntary work

People aged 30-59 carry not only the largest part of responsibility in politics and business, but are also the most active age group in terms of voluntary activity. Demographic change will have an impact here. First because the status of "full" adulthood is achieved significantly later than in former times. Young people often stay with their parents well into their twenties, or at least remain financially dependent on them. Subsequently, stable partnership, marriage and starting a family will occur later in their biography than before. Up to the age of 30 or 35, people remain in a sort of intermediate phase of their lives, a prolonged youth.

Volunteer activity, though, needs reliable and long-term commitment. Most often, people will only be able to make such a commitment when they have established themselves professionally and started a family. Therefore, the growth of so-called "atypical" employment (short-time employment, part-time employment, internships, freelance work) directly reduces the potential for voluntary work, because it hinders long-term relations, both in family and in civil society. This leads to the hypothesis that project- or subject-related short-term forms of civil engagement may become more popular than traditional long-term commitments.

Secondly, demographic change means that more and more people, if they ever become parents, will do so later in their lives. This reduces the potential for volunteer activity in our society. It has often been shown in empirical research that the birth of children triggers community activities and is perhaps the strongest stimulus for volunteer work. With the birth of a child, the feeling of personal responsibility for the future of society grows, and also the will to shape this future. Figure 2 demonstrates that childless people are clearly less active as volunteers than people who have responsibility for a family. [Translation: *couples with pre-school and school age children up to 14 years of age; couples with school age children up to 14 years of age; 3-generation families; couples with children over 15; couples with pre-school children; couples without children, aged 50 and over; all those interviewed; single persons under 50, couples under 50 without children*].

For women, the combination of professional, family and voluntary work forms a considerable challenge and is one of the reasons why they abstain from very time-consuming voluntary responsibilities like local politics. Probably about 20-25 per cent of women and even more men living in Germany today will remain childless all their lives. Most of them live in cities.

Figure 2: Percentage of population engaged in volunteer work differentiated by living-situation (in Germany)

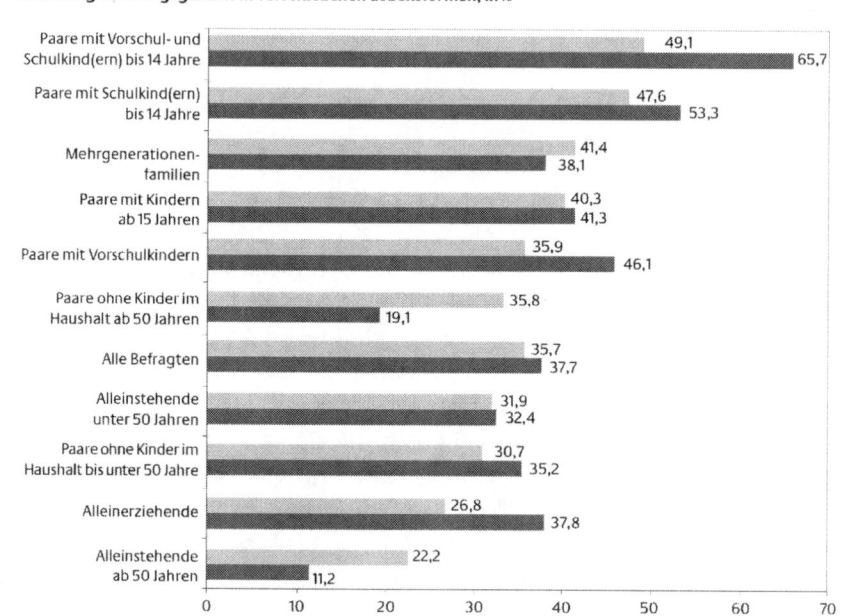

Abbildung 3.3-2: Engagement in verschiedenen Lebensformen, in %

Lebensform		
Paare mit Vorschul- und Schulkind(ern) bis 14 Jahre	49,1	65,7
Paare mit Schulkind(ern) bis 14 Jahre	47,6	53,3
Mehrgenerationen-familien	41,4	38,1
Paare mit Kindern ab 15 Jahren	40,3	41,3
Paare mit Vorschulkindern	35,9	46,1
Paare ohne Kinder im Haushalt ab 50 Jahren	35,8	19,1
Alle Befragten	35,7	37,7
Alleinstehende unter 50 Jahren	31,9	32,4
Paare ohne Kinder im Haushalt bis unter 50 Jahre	30,7	35,2
Alleinerziehende	26,8	37,8
Alleinstehende ab 50 Jahren	22,2	11,2

Source: *Bericht zur Lage und zu den Perspektiven des bürgerschaftlichen Engagements, S.100 (2009)*

Childlessness means that nearly one-third of the adult generation will not automatically have intergenerational relations and links fostering solidarity with young people. This will have different consequences for voluntary work: participation in childcare and school contexts is not relevant for these people, but they might have more time for voluntary work in the evenings and at weekends for trade unions, political parties, and citizens' associations. Furthermore, civil engagement may be an important means of bridging periods of unemployment, and to gain additional qualifications which may lead to a new job.

To conclude, growth of childlessness and late transition to parenthood do not automatically lead to more free time for voluntary work. In all probability, these trends will postpone the biographical phase of intense civil involvement to a later stage in life, thus shortening it. Therefore, to foster civil engagement for adults, it is also necessary to study the labour market and to provide well-paid, stable jobs. This stability is necessary for high levels of volunteer work, whereas unemployment diminishes participation rates. For women, good childcare is also a prerequisite for voluntarism, because otherwise there will just be no time and energy left for activity other than job and family.

3. The older generation and voluntary work

The over-sixties are the only age group that is growing at present, and will soon make up for over 50 per cent of the population. At the same time, the number of healthy and active years of life is increasing. In the near future, a large proportion of people aged 60-85 will no longer be involved in professional activities, but will still be highly productive and powerful. In addition, they will have fewer grandchildren to care for.

Many of those currently aged between 50 and 65 had a new type of participation experience around the reform years of 1968; they are better-off than the cohorts before and after them. The women among them have had the opportunity to emancipate themselves from older role models: they have a driving licence, professional qualifications and varied conceptions of a woman's life. Most of all, the new seniors, "digital immigrants" that they are, have access to the new network and interaction tools of the internet.

We are currently witnessing the emergence of new cultural models for this stage of life. Care for family members, whether the elderly or grandchildren, may still seem dominant in older women's lives, but the outlines of a new picture of old age are already being seen: more leisure-oriented, focused on personal wellbeing, friends, travel, culture and sports. But in times of demographic change, it is a central task of society to develop a third concept, centred on voluntary work in structures outside of family, friendship and neighbourhood. Given that the potential for voluntarism in the younger and the middle age group is actually decreasing, it will be important to activate the potential of voluntary engagement in the third generation. But this will not be an automatic process: the differences between rich and poor, well educated and less well educated seniors are increasing. Voluntary work cannot be seen as an obligation for the elderly, but must remain a free choice alongside other life decisions. The increase in healthy, active years does not automatically mean that seniors in large numbers will stream towards voluntary work agencies.

What we need is rather a different approach. Work is still the main mechanism for integrating people into society. A truly inclusive society cannot tolerate people being excluded from this main integration mechanism just because of their age of retirement. There is an urgent need for other powerful integration mechanisms. Family can be one of these, but is not for everybody. Voluntary work, on the other hand, could be a major way of integrating the elderly (and other excluded groups). It creates participation and gives meaning to life. But this will only happen if we see volunteer work not as a panacea for a shaky welfare state, but as a pillar of social cohesion, as important as public activity and private business.

References

- *Bericht zur Lage und zu den Perspektiven des bürgerschaftlichen Engagements in Deutschland.* Wissenschaftszentrum Berlin für Sozialforschung (WZB). Projektgruppe Zivilengagement Mareike Alscher. Dietmar Dathe. Eckhard Priller (Projektleitung). Berlin: Rudolf Speth, June 2009.
- Gensicke, T., Picot, S., Geiss, S. (2005). *Freiwilliges Engagement in Deutschland 1999–2004. Ergebnisse der repräsentativen Trenderhebung zu Ehrenamt, Freiwilligenarbeit und bürgerschaftlichem Engagement. Durchgeführt im Auftrag des Bundesministeriums für Familie, Senioren, Frauen und Jugend.* München.
- Dienel, C. (2010). *Bürgerengagement und demografischer Wandel.* Bonn: Friedrich-Ebert-Stiftung.

4.4 Family and Education Towards Voluntary Action

Francesco Belletii and Lorenza Rebuzzini
Forum delle Associazioni Familiari

Does the family educate people to develop a pro-social attitude, on which voluntary action is based? In what ways does the family pass on this attitude? What are the factors that thwart this educational dynamic? In 2008 CISF (Centro Internazionale Studi Famiglia/International Centre of Studies on Family) and CSV (Centro Servizi Volontariato/Center for Promoting Volunteering) of Bari, Italy, conducted research on Family and Volunteering, in an attempt to shed light on the connection between family and education for volunteer activity (in Italian, *La famiglia nell'educazione al volontariato/Family and Education for Voluntary Action*).

The research was conducted at local level in the Bari District, in southern Italy. Bari (including the Barletta-Andria-Trani District) is the fifth most populated district in Italy, with almost 1,650,000 inhabitants. According to a previous analysis conducted by CSV Bari, there are over 600 voluntary associations in this district: in most cases they are small associations strongly linked to the territory, that is, the municipality, social services or the local church.

One of the most interesting aspects of this research is the local approach. The family, as well as the voluntary action itself, seem be to strongly determined by the social and economic circumstances of the place in which they are situated. This holds true especially in Italy – a very complex and differentiated country. Living in a wealthy and well organised context seems to lead to different opportunities, as opposed to living in a disaggregated or poor territory. The awareness of the importance of the environment clearly emerges in the research. The second interesting aspect of this research consists in giving voice to those who are volunteering and/or are constantly in touch with families and volunteers: social workers, educators, teachers, members of associations and members of the local government and church. Forty-seven people, aged 30-50, were interviewed or took part in focus groups with a semi-structured interview between February and June 2008. This research is accordingly focused on and enhances the experience-based approach.

1. Theoretical framework

The theoretical framework in which the research was conducted is based on the most recent socio-economic literature on wellbeing. Wellbeing is defined as a condition determined by a complex interrelationship between different factors – to cite the most important ones: wealth, environmental sustainability, social cohesion, and individual freedom. In this context, the concept of social capital is of strategic importance. While we acknowledge the vast debate on the definition of social capital developed in different disciplines such as economics, political science and sociology, in research a "simple" and "common" definition has been used: social capital. Social capital is the social relationships and social networks that are not based on economic or political interest, but that have value for the development of society and economy, and affect the productivity of individuals and groups (Putnam, 2000). Family and volunteering action are environments producing relationships and networks not based on economic or political interests: for this reason, they need to be both recognised as "producers of social capital" and, therefore, as producers of social wellbeing. How can society achieve this?

Family and voluntary action share one fundamental aspect required to build social capital: relationships, in the family as well as in the voluntary action, are not based on the economic model of negotiation but offer an alternative model that can be defined as a 'trustee model'. While the negotiation model is based on maximizing benefit and personal profit, and interaction between individuals is aimed at coming to an agreement, the symbolic codes regulating both family and voluntary actions can be outlined as follows:

- **Gift:** the giving of gifts indicates an interest in the recipient of the gift, and often demonstrates expectation of feedback. At the same time it gives freedom of choice about whether to give feedback or not. The freedom to accept or refuse is instilled in both voluntary action and in family relationships.
- **Reciprocity:** reciprocal actions or relations and mutual exchanges are fundamental social norms and generate innovative human relations. Reciprocity connects past actions with the present, and the present to the future, in its dynamic of receiving and giving back.
- **Trust:** a relationship based on sharing and on satisfied expectations embodies its own reasonableness, even though it may not be perfectly rational. In other words, trust is a reasonable, but not rational, form of behaviour. Nonetheless, economic and social systems are based primarily on trust.

In this framework, families' responsibility to society is not described as a moral obligation and as something added to family from the outside, but as the natural inner dynamic of family life. Family does pro-social action in just "doing family", that is the point. This approach does not seek to be non-judgmental: of course there are families unable to take on this kind of responsibility towards children and society. Nonetheless this dynamic of family life has to be kept in mind when the issue of the relationship between family and society, as well as family and voluntary action is approached.

2. Research in the Bari District

Research in the Bari District highlights a multi-faceted situation. The interviewees claim that their family of origin is a place of positive relationships, where they have learnt and experienced solidarity, care, help and attention to others. They say that they try to build the same positive relationships in their current families, but they find it difficult to do so. As a matter of fact, the Bari District, as a territory, has many problems: unemployment, low salaries, lack of public services (in 2008, the average income in Italy was 18,870 Euros, while in the Bari District, it was 14,830 Euros). Therefore, in the research, family is considered at two different levels:

- pro-social families, the families that enhance and promote volunteering (subject of volunteering);
- vulnerable families, the families that need to be helped (object of volunteering).

3. Difficulties

An in-depth analysis has been conducted on the stress factors thwarting the transmission of pro-social behaviours in today's families in the Bari District – even if it seems that the analysis can be applied to the whole Italian territory, as well as to other countries. Interviewees detected three main points of vulnerability:

- **Social isolation.** Social relations are less frequent and strong, and families are left alone. Therefore, the relationship of mutual trust between families has faded. Moreover, family members are under stress due to the many tasks they have to accomplish: care for children, work, care for themselves, care for other people in the family. Individualism is permeating not only social relations between families, but also relationships within the family.
- **Hedonistic and materialistic values** that are at the forefront of our society push people into constantly looking to "have more", in competition with each other. People need to work to earn more and

more, thus having no time left for others. Media, and especially TV, is seen as the big megaphone through which these values are instilled in people.
- **Adults' vulnerability.** Today's adult generation is seen as being very vulnerable, particularly with regard to the ability to build meaningful relationships and educate and define rules for children.

4. Attitudes to volunteering: an analysis

Interviewees agree that they have learned their attitude to voluntary action in their families of origin, even though debate is open on the ways in which this attitude is transmitted. Nonetheless, they underlined some factors that help families to be pro-social and to instil attitudes leading to voluntary action:

- **Stable relationships.** Interviewees agree that stable relationships are the basis of pro-social action: a stable relationship means openness to others' needs, capacity for dialogue, respect, commitment, support (all these terms have been extracted from the interviews).
- **Dialogue and Rules.** According to interviewees, dialogue is essential "in transmitting rules and values and instilling trust in your children, in your relatives, in persons near you. With dialogue, rules can be set and values can be transmitted peacefully", says a woman involved in volunteering.
- **Lifestyles based on different values.** The pro-social family lives and promotes values based on solidarity; these values are promoted through an "alternative" lifestyle that rejects consumerism.
- **Economic security.** Many interviewees stress that economic security, which means a stable job with a decent salary, is necessary. They claim that only when people have developed a good level of self-confidence can they commit to helping others.
- **Stable policies.** One of the most interesting outcomes of the research is that it shows the need for stable political support for volunteering projects. Often a good project is not undertaken, due to political choices and the way funds are allocated: this leads, of course, to the loss of potentially beneficial outcomes.

5. Conclusions

The research carried out in the Bari District has not produced definite conclusions, but opens up new research questions. In other words, we cannot measure "how much" voluntary action a family can generate, but we can conclude that it is necessary to work on the interrelation and the synergy between family and volunteering. Enforcing the alliance between pro-social families and pro-social action can become a realistic approach to social innovation in the Third and Fourth Sectors[1]. Family associations, which have a self-help and pro-social approach, are a reality that can be reinforced and can help in enhancing pro-social action and in transmitting pro-social values, but further investigation is needed. This can be achieved if the logic of empowerment is applied both to pro-social families and vulnerable families: vulnerable families need not be defined by their needs and their lack of capabilities, but rather by their capacity to "start over again". At the same time, pro-social families need to be supported: many families still feel alone in a society that nowadays seems governed by consumerist and hedonistic values, and they feel challenged in their ability to be real drivers of change.

References

- *La famiglia nell'educazione al volontariato. Costruire insieme capitale umano e sociale,* Bari: Edizioni Pagina. 2009. Available from: *http://www.csvbari.com/opera.html?start=5.*
- Putnam, R. D. (2000). *Bowling Alone: The Collapse and Revival of American Community.* New York: Simon & Schuster.
- Donati, P. (ed.) (2007). *Ri-conoscere la famiglia: quale valore aggiunto per la persona e la società?* Cinisello B.mo (MI): Edizioni San Paolo.

---

[1] The "Third Sector" refers to structured volunteering organizations, while the "Fourth Sector" refers to more informal networks of families and family associations, neighbourhood and help/self-help groups, even though there is still some debate on the definition of the relatively recent Fourth Sector (see a completely different definition at *http://www.fourthsector.net*).

4.5 Towards the European Policy Agenda on Volunteering: Taking Into Account the Needs of Families

Edited by Gabriella Civico
European Project Manager EYV 2011 Alliance

The European Year of Volunteering (EYV) 2011 Alliance is an informal, open and growing group of 35 European networks active in volunteering. The EYV 2011 Alliance was formed in December 2007 with the purpose of starting a common campaign towards establishing the European Year of Volunteering in 2011. Three years later at the start of EYV 2011 this accomplishment stands as a shining example of what can be achieved when Civil Society works together.

Thanks to the financial support offered by the European Commission in the form of a project grant in the Citizenship Programme, and additional co-funding offered by Robert Bosch Stiftung, Fundacion Telefonica and Alliance Steering Group members, the EYV 2011 Alliance has been able to establish a project work programme supported by two full-time staff in its secretariat at the European Volunteer Centre (CEV) in Brussels. As a critical contact point and driving force for the facilitation of the joint effort to lobby for the Year, it was agreed by EYV 2011 Alliance members that CEV should be the grant holder for the EYV 2011 Alliance Project and also therefore host the Secretariat.

Following the White Paper on Youth, the Member States in 2002 recognised volunteering as a key element of youth policy, and a lot of research, exchanges of experiences and discussions have taken place since then. This work has shown its impact over the years and should be enlarged to all age ranges, since only then can policies which take into account all aspects of all generations of families become a real possibility. Whereas various EU activities, programmes and policies tackle volunteering at a European level, these initiatives have so far not taken into account the variety of volunteering in Europe and its horizontal nature. Actions in the field of volunteering at the EU level are restricted to certain age groups and policy areas, and neglect the horizontal nature of volunteering and the relevance of volunteering in solving a variety of political, social and economic challenges that the EU faces today:

- Volunteers are an example of **active civic participation**. They engage in their communities, without financial motivation, for the benefit of other individuals and society as a whole. Volunteering, as an expression of Active European Citizenship, was recognised by the European institutions in a number of areas, especially through the Europe for Citizens' Programme[1].

- Volunteers **put into practice European values of solidarity and diversity**. They are the expression of the EU slogan 'United in Diversity' as these are people of all ages, women and men, employed and unemployed, people from different ethnic backgrounds and belief groups and, finally, citizens of all nationalities. However, those involved do not necessarily make the link between their engagement and European values. People become involved because they feel that they can make a positive contribution to society and because they benefit themselves. It does not occur to them that all over Europe, people engage in volunteering for the same underlying values and motivations. At a time when the EU lacks a link with its citizens and wants to create more ownership of the European project based on solidarity and mutual understanding, it can no longer afford not to contribute to creating the logical link between voluntary engagement at a local level and implementation of European values.

- The Resolution of the European Parliament [A6-0070/2008][2] recognises the **contribution that volunteering makes to the economic and social cohesion** of the European Union. The report says that volunteering makes an important contribution to social integration at the local level, and that it contributes to partnerships which are key for making full use of the European regional and structural funds.

- Volunteers are the **main agents when it comes to social inclusion**, through their engagement with the socially excluded or those at risk of social exclusion. Volunteering is a tool for the empowerment of all, and especially of those that are socially excluded, because it is a means by which citizens can be and feel useful and re-connect with society. This is highlighted in the Youth Pact[3], the White Paper on Youth[4], and in the Bureau of European Policy Advisers' (BEPA) report "Investing in Youth: From Childhood to Adulthood"[5], but it applies equally

---

[1] See *https://secure.wikimedia.org/wikipedia/en/wiki/Europe_for_Citizens*.
[2] See *http://www.europarl.europa.eu/sides/getDoc.do?pubRef=-//EP//TEXT+REPORT+A6-2008-0070+0+DOC+XML+V0//EN*.
[3] See *http://europa.eu/youth/news/index_1794_en.html*.
[4] See *http://ec.europa.eu/youth/glossary/word366_en.htm*.
[5] See *http://ec.europa.eu/dgs/policy_advisers/publications/docs/Investing_in_Youth_25_April_fin.pdf*.

to all age groups. The World Organization of the Scout Movement[6] and its national organisations have carried out a number of projects across Europe on involving and reaching out to minorities and other disadvantaged children (including children with different religious backgrounds, disabilities, Roma, etc.).

- This applies in particular to the **integration of migrants** into our societies. Integration - as a two-way process of mutual accommodation between migrants and the host society - needs tools and instruments that bring people together and which enable them to work on common projects. The many volunteer initiatives and projects in Europe demonstrate the added value that active participation brings in this area. They also demonstrate volunteering as a factor in - and indicator of - the integration of migrants in host communities[7]. Volunteers are also involved in programmes for the integration of migrants (labour orientation, training, etc.).

- **Social services of general interest** in Europe depend largely on the contribution of volunteers. Actions of voluntary organisations implemented in Member States show that volunteers contribute significantly to the services provided in the **health and social care** sector through visiting services for socially isolated people and day centres for older people, people with Alzheimer's disease and homes for children in need of special care, coaching activities which support and empower people to take charge of their own lives again (for example, so-called friendship courses), organising holidays for people with disabilities and/or chronic disease, assistance to drug users and prisoners, assistance to people living with HIV/Aids (care, hotlines, counselling, information), and assistance to women threatened by domestic and/or gender-related violence.

- Volunteering is a means of encouraging **active ageing.** Volunteers not only provide complementary home care for older people (psycho-social support), and organise recreational and sports activities for seniors, but older people themselves who become volunteers stay healthy and active for longer and have opportunities to share their life experience with younger generations. The recent Flash Eurobarometer 247 survey conducted in September 2008[8] shows that

---

[6] See http://www.scout.org/en/information_events/resource_centre/library/reaching_out.
[7] See for example CEV INVOLVE project, http://www.involve-europe.eu (2006).
[8] See http://ec.europa.eu/public_opinion/flash/fl_247_sum_en.pdf.

73 per cent of older respondents indicated that they would consider participating in community and volunteer work after retirement. Forty-four per cent of persons also said that they have already planned or plan to get involved in volunteer work. Volunteering offers great potential for the EU when it comes to active ageing and demographic change. This issue was first raised during the conference on Intergenerational Solidarity for Cohesive and Sustainable Societies during the Slovenian Presidency (27-29 April 2008)[9] as one of the main topics, and later as one of the main topics of the 2nd European Demography Forum held in Brussels in November 2008[10].

- Volunteering plays an important role in **maintaining and restoring family links.** This helps people to find family members they have lost because of wars, conflict and disasters, and supports people who have a missing family member or friend. Volunteers contribute significantly to the enhanced satisfaction of family life and proper work-life balance, as they often engage in areas such as childcare and care of older generations, which are perceived by many Europeans as the main difficulties in family life (Flash Eurobarometer 247, 'Difficulties in daily life faced by families').

- Volunteering contributes to **tolerance, peace building, conflict resolution and reconciliation of divided societies**. The CEV project on this topic (V::I::P/2008[11]), as well as many other projects and activities of the members of the Alliance, have shown that voluntary activities exercised together by the local inhabitants for the benefit of their communities increase people's tolerance and intercultural skills, reduce racism and prejudice, contribute to intercultural and inter-religious dialogue, empower people to be active in the recovery of their communities and connect them with the societies in which they live.

- Voluntary activities are part of **informal and non-formal learning** for people of all ages and at all stages of their lives. Volunteering contributes to personal development and to learning skills and competences, thus **enhancing employability.** Volunteering is accordingly part of the Lisbon strategy towards a more competitive European

[9] See *http://www.eu2008.si/en/Meetings_Calendar/Dates/April/0427_EPSCO.html?tkSuche=ajax&globalDatum=01.02.&multiDatum=29.05.&veranstaltungsart=&globalPolitikbereich=&visiblePath=/htdocs/fr&.*
[10] See *http://ec.europa.eu/social/main.jsp?catId=88&langId=en&eventsId=121.*
[11] See *http://www.cev.be/data/File/VIP_Report_2008.pdf.*

labour market, providing **life-long learning opportunities** that arise when people volunteer. This was confirmed by the Commission Communication 'Making a European Area of Lifelong Learning a Reality'[12], Resolution of the Council on the recognition of the value of non-formal and informal learning within the European youth field [2006/C 168/01][13] and a number of the European Youth Forum's reports and projects, etc.

- **The sports sector** involves the largest numbers of volunteers and participants, and this makes it the largest voluntary, non-governmental organisation activity in Europe. Volunteers are the most important and indispensable resource of sports clubs. According to the European Non-Governmental Sports Organisation (ENGSO), the "labour force" of sports clubs consists of 86 per cent volunteers and only 14 per cent paid staff. The impact of volunteering in sport on EU policies is manifold, and this was recognised in the White Paper on Sport [COM (2007) 391 final][14] and in the Commission Action Plan "Pierre de Coubertin" SEC(2007) 934, Brussels, 11.7.2007[15] which calls for promoting volunteering and active citizenship through sport, and recognises that volunteering reinforces active citizenship and provides many opportunities for non-formal education which need to be recognised and enhanced. Sports NGOs and networks such as ENGSO underline that the EU still needs to work on issues such as taxation (maintain a special tax regime for not-for-profit sports organisations, create additional tax incentives for volunteers, i.e. deductibility of tax from donations), education (design European modules to train volunteers), EU funding programmes (make programmes more accessible for volunteers), and employment (enhanced recognition of voluntary work in sport), etc.

- Volunteers are the **backbone of Europe's civil protection force**. Red Cross and Johanniter International experiences show that volunteers are indispensable for disaster response and preparation activities, for first aid services, and education, as well as in relief exercises i.e. ambulances, first aid, psycho-social support and emergency responses.

- **Development policies** are practically impossible to implement without the contributions of volunteers. Volunteers engage in humanitarian

---

[12] See *http://www.bologna-berlin2003.de/pdf/MitteilungEng.pdf*.
[13] See *http://www.europass-ro.ro/doc/resolution.pdf*.
[14] See *http://ec.europa.eu/sport/white-paper/doc/wp_on_sport_en.pdf*.
[15] See *http://ec.europa.eu/sport/white-paper/doc/sec934_en.pdf*.

missions and provide assistance to refugees (including humanitarian assistance, reception centres including legal counselling, health services, mental health care, etc.). More than 6,000 volunteers engage every year with United Nations missions alone. Furthermore, Article 188 of the Lisbon Treaty decrees that the EU 'establish a framework for joint contributions from young Europeans to the humanitarian aid operations of the Union' in the shape of a European Voluntary Humanitarian Aid Corps.

- **Corporate (employee) volunteering** schemes are increasingly seen by European companies as a means of connecting to society, investing time and resources in their communities and giving concrete meaning to their CSR policies. Thousands of individuals across Europe are already benefiting from the help and support offered by employee volunteers. The European Parliament Resolution of 13 March 2007 on 'Corporate social responsibility: a new partnership' (2006/2133(INI)) underlines the importance of projects involving employee community engagement and calls on the Commission to fulfil its commitment to developing policies that encourage the staff of EU institutions to undertake voluntary community engagement. In the Communication from the Commission on Implementing the Partnership For Growth and Jobs: Making Europe a Pole of Excellence on Corporate Social Responsibility COM (2006) 136 final[16], the European Commission commits to step up its policy of promoting the voluntary and innovative efforts of companies on corporate social responsibility (CSR). This still seems to be unfulfilled. The business platform ENGAGE, in its publication CSR Laboratories: Bringing the European Alliance on CSR to Life[17], demonstrates the effectiveness of employee community engagement in improving the skills essential for employment amongst disadvantaged and socially excluded groups of people within the EU. It urged the European Commission to support and encourage employee volunteering by announcing the European Year of Volunteering 2011.

- **Finally, volunteering is economically important.** The research of the Institute for Volunteering Research, 'Volunteering Works - Volunteering and social policy'[18], shows that for every Euro that organisa-

---

[16] See *http://eur-lex.europa.eu/LexUriServ/LexUriServ.do?uri=COM:2006:0136:FIN:en:PDF*.
[17] See *http://www.pr.org.rs/upload/documents/ENGAGE per cent20Skills per cent20for per cent20Employability per cent20Report.pdf*.
[18] See *http://www.ivr.org.uk/evidence-bank/evidence-pages/Volunteering+Works+-Volunteering+and+social+policy*.

tions spend on supporting volunteers, they receive an average return of between three and eight Euros. Moreover, the Comparative Non-Profit Sector project at Johns Hopkins University (USA) revealed that the voluntary sector contributes an estimated 2 to 7 per cent to the GDP of our national economies. In the UK in 2007 volunteering contributed 48 billion pounds sterling to the national economy (according to the Volunteering England's figures). We do not have the necessary data and instruments in place to properly analyse the economic value nor to raise policy-makers' or funders' awareness of the contribution that volunteers make to our economies.

Since 2007 EYV 2011 Alliance members have been working together to lobby for the EYV 2011, working with other stakeholders and policy makers across the EU. The experience of civil society organisations showed that there was a need at all levels - EU, national, regional and local - to increase volunteering and the awareness of the added value it brings to European society, to celebrate volunteering, involve more volunteers, and to improve the policy framework on volunteering including that related to volunteering and families. There is no Europe without volunteers: they contribute both to its growth, and to its social character. EYV 2011 Alliance members thought that EYV 2011 would increase recognition of this and increase the capacity of volunteering organisations to deliver their missions.

The EYV 2011 Alliance Project Work Plan addresses several critical issues, especially those which are to be discussed within the policy dialogue to take place during the Year. These will draw upon the widespread expertise gathered together in the 100 members of the EYV 2011 Alliance Working Groups, many of whom are volunteers themselves.

Working groups have been established on six different themes (Quality of Volunteering, Legal Framework of Volunteering, Volunteering Infrastructure, Recognising Volunteering, the Value of Volunteering & Employee Volunteering). The groups met for the first time on 7 and 8 January 2011 in Budapest. The working group members have been nominated by EYV 2011 Alliance member organisations and will meet at least five times during 2011, including during the Kick-Off Conference (already held in Budapest) and three further Working Group meetings to be held in Brussels in March, May and September 2011. The EYV 2011 Alliance Working Group closing conference in Poland (December 2011) will approve the "EYV 2011 Alliance European Policy Agenda on Volunteering", which will then be presented to policy makers at the general EC EYV 2011 closing conference, to be held in Poland later in December 2011.

Working towards the European Policy Agenda on Volunteering the EYV 2011 Alliance will address:

Quality

- Work towards a common understanding of "quality volunteering".
- Clarify the roles and responsibilities of the organisers of volunteering in ensuring quality volunteering experiences.
- Identify and disseminate good practice in the field of quality assurance and quality assessment tools used by volunteer organisations.

Legal framework

- Map research on the legal status of volunteers in Europe.
- Collect concerns in terms of legal barriers and bottlenecks caused by legal provisions at any level that result in limiting volunteering in Europe.
- Advocate recommendations for improvement of the legal status of volunteers and a clear legal status for volunteers everywhere in Europe.

Volunteering infrastructure

- Extract good practice indicators of an enabling volunteering infrastructure at different levels, feeding into a European framework recommendation that allows for the national diversity of volunteering to be respected.
- Identify key legal features for volunteer organisations, so as to provide an enabling infrastructure.

Recognition

- Map the existing tools for recognition of volunteers - extract good practice examples.
- Map tools of how to recognise the contribution of volunteer organisations.
- Formulate recommendations for better recognition of volunteering in different areas and by different tools.
- Devise a strategy for implementation of the recommendations.

Value

- Identification of tools and ways to identify, measure and express the value of volunteering.
- Valuing volunteering as an important creator of human and social capital, cohesion and wellbeing, encompassing the provision of services and effective interventions where other policies may fail.

- Valuing the contribution of volunteering in positively shaping the European society.
- Valuing volunteering as an expression of solidarity, a value which is not only in great need in the current economic and social climate, but also one upon which the EU has been built.
- Recognising the contributions of volunteers as match-funding in all European and national project funding.

Employee volunteering

- Increase understanding of the concept of Employee Volunteering as a key element of Corporate Social Responsibility (CSR).
- Achieve recognition of Employee Volunteering as a means and key component of putting CSR concept into practice.
- Contribute to making Employee Volunteering accessible for all, in all sectors (private, public and non-profit).

All in all, EYV 2011 will contribute to helping the wider public understand why volunteering is a critical issue across the EU, and this will include how it impacts on families and family life.

Volunteering is freely given, but is certainly not cost-free, and the EYV 2011 Alliance members believe that volunteering and volunteering organisations need and deserve targeted support from all stakeholders – volunteer organisations, government at all levels and businesses. The EYV 2011 Alliance Working groups will spend EYV 2011 working towards this vision of an enabling volunteering environment including a volunteering infrastructure that takes the realities and needs of families across the EU into account. EYV 2011 Alliance members are committed to engaging with the EYV 2011 together with key EU stakeholders, especially the European Commission. They wish to demonstrate and showcase what good policymaker-civil society partnerships can achieve, in particular through presentation of the European Policy Agenda on Volunteering, which will be delivered before the end of 2011.

4.6 Volunteering and Service in the United States

Barb Quaintance
AARP, USA

The United States has a strong tradition of volunteering and service. Of course, families, neighbours and co-workers support one another in times of need. An unprecedented number of Americans, however, also work to help strangers. And today, both kinds of service are stronger than ever.

Many non-profit organisations engage volunteers in service – often to help youth get a good education, assist with economic needs, improve the environment, support health and ageing services, and assist with disaster preparedness/relief. Some of the larger faith-based domestic charities include Catholic Charities USA, Lutheran Social Services, and Volunteers of America. Some of the larger secular volunteer organisations include Red Cross, AARP, Points of Light Institute, HandsOn Network, and United Way.

Even with this strong history, today is a new day for volunteering and service in the United States. More people are volunteering, service is now seen as a solution to key challenges, and individuals and families are enjoying clearer benefits – whether as volunteers, as beneficiaries or both.

Why is service in America on the rise today? National leadership, greater opportunity, and a clearer effort to weave service into our schools, workplaces, and national calendar of holidays and anniversaries are critical factors. Today, volunteer work is seen as a critical strategy to solve some of the toughest challenges experienced by individuals and families. Below, we take a closer look at each of these factors – and the impact on families who serve, as well as families who are served.

1. Presidential leadership

Presidential leadership in recent years has been critical in strengthening volunteerism in America. Fifty years ago, President Kennedy challenged Americans to serve with his famous quote, "ask not what your country can do for you; ask what you can do for your country". Kennedy went on to initiate The Peace Corps programme. Next, President Johnson created Volunteers in Service to America (VISTA) and the National Civilian Community Corps. The seeds of organised, widespread volunteerism were planted.

In his 1989 Inaugural Address, President George H.W. Bush described, "a thousand points of light, all the community organisations that are spread like stars throughout the nation, doing good". He acted to bring this optimistic vision to fruition in signing the first national service act. His efforts also led to the development of the Points of Light Foundation – which ultimately grew into Points of Light Institute, the largest volunteer management and civic engagement organisation in the US.

In 1993, President Bill Clinton signed legislation to align national service programmes under a new government entity (Corporation for National and Community Service) with the addition of a new programme, AmeriCorps – which enables individuals to serve one year to help strengthen a community. In swearing in the first class of 20,000 AmeriCorps members, Clinton said, "Service is a spark to rekindle the spirit of democracy in an age of uncertainty. When it is all said and done, it comes down to three simple questions: What is right? What is wrong? And what are we going to do about it? Today you are doing what is right – turning your words into deeds".

Building on the culture of giving and service following September 11, 2001, President George W. Bush expanded and promoted volunteer service by creating USA Freedom Corps, to co-ordinate volunteer efforts and help all Americans find opportunities to serve. He also initiated the President's Volunteer Service Award programme to recognise volunteers who contribute substantial time in service to others.

Upon taking office, President Barack Obama further strengthened the nation's commitment to national, organised service. After calling for legislation to expand the ability of service to solve challenges in communities, he signed the Edward M. Kennedy Serve America Act, which further expands volunteer opportunities and seeks to identify and take to scale volunteer programmes with proven results. Obama also called for a "Craigslist" for service, getting Craig Newmark himself to help build that[1]. Finally, President Obama continues to set a powerful example for family volunteering as he volunteers several times each year with his own family at schools, food pantries and other sites needing help.

2. More accessible volunteer opportunities

Americans have a hunger for volunteer work. Families have traditionally heard about volunteer opportunities through faith organisations, schools and community groups, but many eager volunteers have had difficulty finding suitable volunteer opportunities. Others may already be engaged, but are eager to do more. Therefore, several groups have been working to make service more accessible.

As part of that, AARP *Create The Good* is providing a network of opportunities for people to get connected to make a positive impact. *Create The Good* offers the new online database of opportunities searchable by locality; the ability

---

[1] "Craigslist" is a centralised online network connecting people with jobs, housing, items for sale, etc.

for anyone to post an opportunity to request more volunteer help, and simple, fun project ideas people can use to develop their own volunteer efforts. These resources also are available at compartiresvivir.org and similar websites.

The idea is to enable individuals and families to use the database to find local volunteer opportunities that fit their time, skills and passion. Or if they prefer, they can use one of the many available 'how-to guides' to design a convenient project that works for everyone in the family. As we say as part of AARP's *Create The Good* initiative, people can make a difference in five minutes, five hours or five days.

One example of such self-initiated service is that of New York resident Olga El Sehamy. Olga recalled that her mother often offered food to street vendors in Mexico, and that her husband lacked housing when he came to the United States. She had a strong desire to give back on a national day of service, but her work schedule precluded volunteering with an organised group. So, she recruited her family to help provide food to people living in the streets and subway stations. She and her husband prepared dozens of meals, and her son and his college friends distributed the food throughout the morning. The college students felt so good about the project that they asked if they could do it again soon – and they did!

This kind of flexible volunteering is increasingly popular. Fifty-seven per cent of Americans over the age of 45 reported that they engaged in service efforts they organised themselves in 2009, up from 34 per cent just six years earlier. Another survey revealed that the number of people who worked with their neighbours to fix a community problem rose from 15.2 million in 2007 to 19.9 million in 2008.

3. The growing ethos of volunteering

Volunteerism has long been a part of American culture. The last decade, however, has embedded it far more deeply into the American psyche.

First, service is on our calendar. We now have two national days of service – Dr. Martin Luther King, Jr. Day of Service and 9/11 Day of Service and Remembrance. These offer regular and visible chances for families to help others on a possible shared day off. Further, many families are beginning to celebrate Veterans Day by engaging in service with and for veterans. There's also National Volunteer Week in April, which includes a Youth Service challenge. And the fourth Saturday in October is Make A Difference Day.

Second, service is fully integrated into our system of education through student service learning. Through these programmes, students are asked to engage in service projects with their teacher and classmates and also help others in the community outside of school hours – on their own or with family. In practice, many younger students volunteer with family members;

as they grow older they begin to volunteer with their peers. Often, students are asked to play a role in selecting the service project and reflecting on what it means to them. These programmes begin as early as kindergarten and run through high school – sometimes as a graduation requirement. Most colleges also offer service learning opportunities.

At their best, student service learning programmes empower students to learn skills, develop leadership, solve problems and help others. Teachers increasingly integrate service with curricula – for example, learning how chemistry might be used to clean up a river. Additionally, these programs aim to develop a love of service, setting the stage for successful family volunteering. A recent focus group conducted among teens and 20-somethings in the state of Maine revealed a lasting positive impression about service. Participants said they volunteer for many reasons, including self-improvement, peer pressure, a desire to give back, because they've been involved from a young age, a desire for new experiences and...for fun!

Third, there has been a boom in volunteering in the workplace. Many employers find they are more likely to attract and retain talented staff if they include effective service opportunities as part of the workplace environment. And many corporations are including engagement and support of service as part of their overall marketing plan, e.g. in advertisements, in stores and at sporting events.

Finally, service is increasingly evident in a growing number of cities. A new "Cities of Service" initiative was introduced in 2008 by New York City Mayor Michael Bloomberg. Already it includes over 100 major cities, and it is spreading all across America. Mayors are taking a leadership role, bringing communities together to identify which challenges to address, and deciding how they can leverage assets to achieve strong impacts together. While this initiative is just beginning to come to fruition, there are already innovative and effective efforts underway to support youth in staying in school, to plant and maintain community gardens far and wide, to empower an entire city to learn emergency response skills, and much more.

So, whether in school, at work, on holiday or attending a football game, American families are surrounded by opportunities to help others. As they engage, they benefit from the opportunity and the reward – whether they are in a giving or taking position any given day.

4. More people are engaged in service and benefiting from volunteerism

Wealthy or not, young or old, single or married, across all cultural and ethnic groups, Americans are stepping forward to make things better for individuals and communities. The growth in volunteering reported in 2009 represented the greatest annual increase since 2003. A government survey capturing a broad definition of service revealed that 58 per cent of Americans helped a neighbour at least once a month in 2009. Even people facing challenges of their own are helping, often verbalising that they know how tough it is to have even less.

Families are very much a part of this trend. Married people, especially married women, have some of the highest rates of volunteering around. And, rates of volunteering are up for all population groups, including children.

What are the effects on families or anyone who volunteers? Numerous research studies have revealed that volunteers tend to be happier and enjoy lower levels of stress, lower incidence of depression and higher self-esteem.

Yet, while volunteerism is booming, it hasn't yet grown to full capacity. Tens of millions of Americans want to volunteer more than they do today. And, of course, millions of families need more help. This conundrum exists because many interested volunteers don't know how to find their best role – and at the same time, volunteer organisations lack capacity to fully engage all those who want to help more.

Renewed efforts to make volunteering more accessible should give more Americans access to the benefits of "doing good". And, with that increased capacity to reach out to others, we hope to solve more of the problems we face as a society.

5. Measuring the impact of service in improving our world

It seems fair to assume that increases in volunteerism will directly improve our society. More youth mentors, more people helping prepare for disasters, and more soup kitchen volunteers should mean a better world. But it's also fair to assume that some service programmes are more effective than others.

Today there is a serious effort to identify the service programmes that make the greatest difference. While many programmes already have proven effectiveness, others lack demonstrated results. We think they work well, but we don't know. Organisations interested in improving service are today finding ways to measure the impact of nearly every volunteer effort around (not always easy), identify those that are most effective, and to find the resources to expand the most effective initiatives to a size that enables them to solve key challenges.

For example, what is the best way to help young people? Data demonstrates that mentoring and tutoring programmes are effective, so increasing the number of volunteers in programs such as Big Brothers/Big Sisters, Experience Corps, JumpStart or City Year makes sense. But with so many high schools dubbed dropout factories, some communities are testing whether it's also helpful to engage neighbourhood adults in knocking on the doors of students who don't arrive at school each morning.

Or take a look at emergency preparedness: the American Red Cross and other organisations train and deploy volunteers in the event of disaster – to help supplement the good efforts of rescue personnel. These are effective programs, but are there other ways communities should prepare for emergencies? For example, New York City service leaders are inspiring thousands of residents to learn first aid from local firemen, and will measure and test whether this is a valid way to help prepare for emergency situations.

The volunteer component is part of identifying and deploying strategies with the strongest impact. It's critical to identify, recruit and retain the right volunteers for each effort. Depending on the situation, the most effective volunteer might be a highly skilled individual who can contribute at least ten hours per week, or it might be a family who lives in the neighbourhood who can help out ten minutes a day.

In 2011, helping others is increasingly in the mainstream of American life. Families and individuals have more ways - and more effective ways than ever - to volunteer their time to help solve pressing societal challenges. And we're aware that in "doing good" we're also improving our own lives: in addition to the warm feeling that comes from helping others, families also can build enjoyable memories together.

Contributors

Ursula Adam

Ursula Adam has been a student of sociology at the Otto-Friedrich-University of Bamberg since 2005. From 2007 until 2008, she was at the Radboud Universiteit in Nijmegen, supported by the Erasmus programme. She worked for a period as an assistant at the Institute for Family Research at University Bamberg (ifb) in a state-funded project on demographic change. Currently she is a visitor at the IMES-institute in Amsterdam. Her main interests include methods of empirical research, gender and family research, migration as well as international comparison.

Loreen Beier

Loreen Beier (Dipl. Soz.) studied sociology at the Otto-Friedrich-University of Bamberg, from where she graduated in April 2008. During her course of studies she focused on empirical research methods and statistics, as well as demography. Her research interests include quantitative and qualitative research methods, comparative analyses of welfare states and their effects on families, especially gender issues and the compatibility of work and family, with specific attention to particular differences between the new and old Länder in Germany.

Francesco Belletti

Francesco Belletti is a sociologist and is currently President of Forum delle Associazioni Familiari (FDAF), an Italian umbrella NGO representing 49 family associations and 20 regional forums. He is also Director at CISF – Centro Internazionale Studi Famiglia (International Center for Family Studies), a research centre based in Milan, and editor of a bi-annual report on Italian families. He has published many articles and books on family life and local family policies projects.

Julie de Bergeyck

Julie de Bergeyck joined the MMM team to work on FAMILYPLATFORM as Project Manager. She is a mother of three and has a background in communications. She worked in the internet advertising business in Brussels and in the US where she spent eight years in a leading advertising agency, before working at Microsoft in Brussels. She recently took a three-year break to raise her third child and has volunteered for different local organisations.

Herwig Birg

Herwig Birg is Professor Emeritus for demographics and long-time Director of the Institut für Bevölkerungsforschung und Sozialpolitik (Institute for Population Research and Social Politics) at the University of Bielefeld. He has published many books on demographic change.

Zsuzsa Blaskó

Zsuzsa Blaskó studied Economics and Sociology and did her Ph.D in Sociology. She has been a research fellow at DRI since 2002. Her research interests include social inequalities, gender issues and family policy – with a special focus on families with young children.

Gabriella Civico

Gabriella is Project Manager of the European Year of Volunteering 2011 Alliance. She has a degree in Social Policy and Education from Surrey University (UK) and a Masters in Education in E-learning from the University of Hull (UK). Her professional background is in business and the youth NGO sector where she has worked as an administrator and also as a trainer and expert. Through her work in the youth field she was involved in the campaign to establish 2011 as the European Year of Volunteering and has been a volunteer in a variety of fields since childhood.

Paul Demeny

Paul Demeny has been Distinguished Scholar at the Population Council since 1989. He has served as the editor of the Council's Population and Development Review, which he founded, since 1975. His research focuses on population policy, international migration, and replacement fertility issues. Among Demeny's professional affiliations are the American Association for the Advancement of Science, where he has been a Fellow since 1974; the Population Association of America, whose president he was in 1986; and the International Union for the Scientific Study of Population, which named him Laureate in 2003. In addition, he was recipient of the 2003 Olivia Schieffelin Nordberg Award for Excellence in Writing and Editing in the Population Sciences and is now an External Member of the Hungarian Academy of Sciences. Invited lectures and professional engagements have taken Demeny overseas more than 180 times during the past four decades and he has been a consultee for many international organisations including the World Bank, the United Nations, and the National Academy of Sciences.

Christiane Dienel

Prof. Dr. Christiane Dienel is Scientific Director of the Nexus Institute for Co-operation Management and Interdisciplinary Research. From 2006-2009, she was Secretary of State in the Ministry for Health and Social Affairs in Magdeburg, Saxony-Anhalt. She teaches Social Policy at the University of Applied Sciences Magdeburg. Her research concentrates on demography, family policy, social policy and citizens' participation in Germany and Europe.

Claudine Attias Donfut

Dr Claudine Attias Donfut is a sociologist and the Director of the Ageing Research Department of the CNAV (National Social Security Pension Fund) in France. She also works as Associate Senior Researcher at Centre Edgar Morin, EHESS (Ecole des Hautes Etudes en Sciences Sociales). Her research interests are in social relations and economic transfers between generations, family and the welfare state, sociology of the life course, transition to retirement and ageing, ageing in developing countries, European comparative studies, and the sociology of immigration.

Carla Facchini

Carla Facchini is Professor of Sociology of the Family at the University of Milan-Bicocca. Her research interests include the condition of the elderly and changes in family structures and relations.

Linden Farrer

Linden Farrer joined COFACE (the Confederation of Family Organisations in the European Union) to work on FAMILYPLATFORM in 2009. Educated at Oxford Brookes where he obtained a BA (Hons) in History and Anthropology, and at the University of Sussex where he obtained an MSc in Social Research Methods, he has experience working on projects funded by the European Social Fund, Youth in Action, and the Seventh Framework Programme.

Leeni Hansson

Leeni Hansson, Ph.D, is Senior Research Fellow at the Institute of International and Social Studies (IISS) at Tallinn University, Estonia. She has participated in several research projects funded by the European Commission, Nordic Council of Ministers and Estonian Science Foundation in the area of gender studies, family studies and female employment studies.

Dirk Hofaecker

Dirk Hofäcker, Dr. rer. pol., studied Sociology and Political Economics at the University of Bielefeld, Germany. During his years of study, he spent a research term in 1997 at the State University of St. Petersburg, Russia, financially supported by a TEMPUS/TACIS scholarship. His major research interests include comparative analyses of welfare states, the effects of welfare systems on the social structures of families and labour markets, and quantitative multivariate methods of social science data analysis.

Kimmo Jokinen

Dr. Kimmo Jokinen is the head of the Family Research Centre at the University of Jyväskylä. He was educated at the University of Jyväskylä, where he obtained a Ph.D degree in 1997. His research interests have focused on family studies (recently complex and challenging family relations and emotional security in these relations), cultural studies, children and the media, and school and youth.

Zsuzsanna Kormosné-Debreceni

Zsuzsanna Kormosné-Debreceni is Social Policy Officer at the National Association of Large Families in Hungary (which has 14,000 member families) and Vice-President of the European network FEFAF (Fédération Européenne de Femmes Actives au Foyer). Having been a member of the Hungarian Council for the Equal Opportunities of Women and Men for more than ten years (and a mother of five children) she is a well-known promoter of work-life balance.

Teppo Kröger

Dr. Teppo Kröger is a social policy and social work researcher who has been involved in a number of Nordic, European and global research projects. His research has focused on comparative study of work-family reconciliation and social care systems, covering childcare as well as care services for older and disabled people.

Marjo Kuronen

Dr. Marjo Kuronen is a Senior Lecturer in Social Work at the University of Jyväskylä, Department of Sciences and Philosophy. She was awarded her Ph.D at the University of Stirling (UK) in 1999. Her research interests include gendered practices of parenting and post-divorce family relations, feminist issues social work, and qualitative cross-cultural research methodology.

William Lay

William Lay is an economist and a post-graduate of the College of Europe in Bruges. He has professional experience working in a large multinational companies such as Unilever, and in different European organisations such as Beuc (European Consumers Bureau) and Comitextil (European textile industry lobby). He has been Director of COFACE (the Confederation of Family Organisations in the EU) since 1983 and has a wide experience of NGOs and of the EU institutions.

Carmen Leccardi

Carmen Leccardi is Professor of Cultural Sociology at the University of Milan-Bicocca, where she has been appointed by the Rector as Scientific Coordinator for Gender Issues. She has researched extensively in the field of youth cultures, cultural change, gender and time. Co-editor (1999-2009) of the Sage journal Time & Society, and now Consulting Editor, she was Vice-President for Europe of the International Sociological Association, Research Committee 'Sociology of Youth' (2006-2010). Her recent books include Sociologie del tempo (Sociologies of Time), Laterza (2009); A New Youth? Young People, Generations and Family Life (Editor, with Elisabetta Ruspini), Ashgate (2006); Sociologia della vita quotidiana (Sociology of Everyday Life) (with Paolo Jedlowski), il Mulino (2003).

Sara Lesina

Sara obtained a Bachelor's degree in Communications and Journalism from the University of Torino, Italy, and a Master's degree in Multimedia and Mass Communications from the University of Torino and the University of Helsinki, Finland. Sara is currently a Trainee at the European Commission, DG Communication, working in the Task Force for the European Year of Volunteering 2011. Before that she earned the fellowship "Master dei Talenti 2009" from Fondazione CRT, Torino, to do a one-year internship at the European Foundation Centre in Brussels, working in the Information, Knowledge and Communications Department. Sara is part of the BETA e.V. – Bringing Europeans Together association, Mainz, Germany, where she works as Head of the Translation Services for the organisation of the Model European Union in Strasbourg.

Anne-Claire de Liedekerke

Anne-Claire de Liedekerke, has been president of MMM Europe for nearly three years. Its mission is to represent the voice of mothers to European institutions and to raise awareness of the importance of mothers' role in the social, cultural, and economic development of our societies. She is an art historian and a mother of three grown-up children. Her family has lived in many different parts of the world, which has given her the opportunity to gain experience as a volunteer, with professional commitments. She also launched the yearly guide "Expats in Brussels", of which she was until recently Co-editor.

Ariela Lowenstein

Ariela Lowenstein is Full Professor at the Department of Gerontology, and Director of the Center for Research & Study of Aging at the Faculty of Welfare and Health Sciences at University of Haifa; within the Center she collaborates closely with scholars from Israel and internationally. She is also Head of the Department of Health Services Management and Head of Research at the Max Stern Yezreel Academic College in Israel, and Chair of a National Advisory Committee to the Israeli Ministry of Senior Citizens.

John Macdonald

After brief periods working in the (then) Scottish Office in Edinburgh and in the Cabinet of Sir Leon Brittan QC, Vice-President of the European Commission, John was recruited as an official by the European Commission in 1995. Here he worked in the Directorates-General for Trade, Economic & Financial Affairs, and Education & Culture. More recently, he was the Spokesperson for Education, Training, Culture & Youth with the Slovak Commissioner, Ján Figel'. John began his newest mission, as Head of the European Commission's Task Force for the European Year of Volunteering 2011, in February 2010. John holds a bachelor's degree in Economics from the University of Cambridge, which he complemented with subsequent studies in Law at Cambridge and Heidelberg, Germany. His studies in Germany culminated with a Master's Degree in European Economic Studies at the Europe Institute of the University of the Saarland.

Elisa Marchese

Elisa Marchese, Dipl. Soz., studied Sociology at the Otto-Friedrich-University of Bamberg, from where she graduated in 2009. Her research and work interests include both empirical and theoretical issues, especially in the sociology of family and work, in gender and intercultural studies, youth and welfare services and optimisation of organisations, as well as quantitative and qualitative research methods.

Livia Sz. Oláh

Livia Sz. Oláh is Associate Professor and Lecturer in Demography. Her main research interest is family demography, with a focus on the impact of public policies and gender relations on childbearing, partnership formation and disruption in Europe, especially Sweden and Central-Eastern Europe. In the years 2003-2008 she had a postdoctoral fellowship for studies of gender relations and family behaviour from the Swedish Council for Working Life and Social Research (FAS). She is the co-ordinator of a European research network on gender and contemporary family dynamics, also supported by FAS.

Miriam Perego

Miriam Perego, Ph.D in Sociology and Methodology for Social Research at the University of Milan-Bicocca, carries out research in the field of gender, social change and youth cultures. Within FAMILYPLATFORM, she is working as a Research Assistant on the changing processes in family life in Europe.

Barb Quaintance

Barb Quaintance, graduate of University of Southern California and the University of Illinois, is the Senior Vice President of the Office of Volunteer and Civic Engagement at AARP. This newly created office will help redefine volunteerism by integrating flexibility with challenging opportunities that fit with the nature of members' lives, making volunteering easier and more accessible. Barb has a rich history of collaborating with volunteers, from programmes to state offices, and was instrumental in creating the Annual Day of Service and recently, successful presence at the ServiceNation volunteer summit in September 2008. Barb established AARP's Medicare/Medicaid Assistance Program, an early model of what ultimately grew into the State Health Insurance Programs (SHIP).

Marita Rampazi

Marita Rampazi is Professor of General Sociology at the University of Pavia. Her research interests include social uncertainty, time and identity in reference to the condition of young people in contemporary societies and transformations in family relations.

Lorenza Rebuzzini

Lorenza Rebuzzini works for Forum delle Associazioni Familiari. She graduated in Philosophy and collaborates as a researcher with CISF (Centro Internazionale Studi Famiglia – International Center for Family Studies in Italy). She has published research on widowhood in Italy, in collaboration with Francesco Belletti and also writes for the Italian review Famiglia Oggi.

Epp Reiska

Epp Reiska (BA) is a Master's student in Sociology at the Institute for International and Social Studies of Tallinn University. She has previously participated in projects in the fields of teacher education (annual monitoring of the Induction Year Programme since year 2006) and youth work (assessing the quality of services and opportunities at youth centres from the perspective of both users and providers).

Marina Rupp

Marina Rupp, Dr. rer. pol., studied Sociology at the University of Bamberg, Germany. Her issues are wide-ranging, including questions related to pregnancy, the paths to parenthood, divorce, as well as gay and lesbian families and large families.

Raul Sanchez

Raul Sanchez is Director of the "Institut d'Estudis Superiors de la Família" (IESF) (Institute of Advanced Family Studies), Universitat Internacional de Catalunya (Barcelona) and General Secretary of the European Large Families Confederation (ELFAC).

Birgit Sittermann

Birgit Sittermann, M.A. works as Research Officer for the Observatory for Sociopolitical Developments in Europe at the Institute for Social Work and Social Education in Frankfurt (Main), Germany. She studied politics, and her current work focuses on comparative European Social Policy and Volunteering.

Zsolt Spéder

Zsolt Spéder was educated as an economist at the Budapest (former Karl Marx) University of Economic Sciences, studied sociology in Germany, and in 2001 earned his Ph.D in Sociology. He is currently Director of the Demographic Rresearch Institute, and has taken part in several international comparative projects. His research interests are in family and fertility issues, life-course studies and longitudinal panel research methods. He is Consortium Board Member at UNECE "Generation and Gender Program", and member of the editorial board of the journal International Sociology.

Joan Stevens

Joan Stevens has been Secretary General of MMM Europe for over four years. She has lived in Europe for 17 years and is the mother of five and grandmother of 16 children. Joan is a professional music teacher, and also served as Chairman of a Foster Care Citizen Review Board. She has long experience with volunteer organisations and causes.